Theory and Design for Mechanical Measurements

Theory and Design for Mechanical Measurements

Editor

Bianca Lupei

Theory and Design for Mechanical Measurements

Edited by **Bianca Lupei**

Printed in 2017

ISBN: 978-1-68117-176-0
Library of Congress Control Number: 2015951148

© 2016 by
SCITUS Academics LLC,
616, Corporate Way, Suite 2, 4766,
Valley Cottage, NY 10989

www.scitusacademics.com

Preface

Measurement is the process of comparing unknown magnitude of certain parameter with the known predefined standard of that parameter. Measurements are one of vital parts of not only mechanical engineering but all types of engineering fields. Every branch of engineering comprises two processes: design, and operations and maintenance. The design may be machine design, building design, circuit design, transportation design, and automobile design etc. The operations part includes operation of the machines, automobiles, various plants, circuits etc.

Both, the design, and operations and maintenance involve measurements. For instance while designing automobile we have to consider dimensions of various parts of the automobiles, the loads they can pick up etc. Likewise during the operations of the plant, say like industrial refrigeration plant, we have to measure parameters like pressure, temperature, etc. In the power plant we have to measure various quantities of the coal, the quantity of water in the boiler, the amount of steam produced along with its flow rate, temperature and pressure, the amount of power produced, the outlet temperature of the steam from condenser etc. In the large chemical plants large numbers of such parameters have to be measured.

Theory and Design for Mechanical Measurements provides a timely and in-depth reference to the theory of engineering measurements, measurement system performance, and instrumentation.

Table of Contents

CHAPTER 1

Foundations of Measurement Fractal Theory for the Fracture Mechanic

Lucas MáximoAlves

GTEME – Grupo de Termodinâmica, Mecânica e Eletrônica dos Materiais, Departamento de Engenharia de Materiais, Setor de CiênciasAgrárias e de Tecnologia, UniversidadeEstadual de Ponta Grossa, Brazil

INTRODUCTION

A wide variety of natural objects can be described mathematically using fractal geometry as, for example, contours of clouds, coastlines, turbulence in fluids, fracture surfaces, or rugged surfaces in contact, rocks, and so on. None of them is a real fractal, fractal characteristics disappear if an object is viewed at a scale sufficiently small. However, for a wide range of scales the natural objects look very much like fractals, in which case they can be considered fractal. There are no true fractals in nature and there are no real straight lines or circles too. Clearly, fractal models are better approximations of real objects that are straight lines or circles. If the classical Euclidean geometry is considered as a first approximation to irregular lines, planes and volumes, apparently flat on natural objects the fractal geometry is a more rigorous level of approximation. Fractal geometry provides a new scientific way of thinking about natural phenomena. According to Mandelbrot [1], a fractal is a set whose fractional dimension (Hausdorff-Besicovitch dimension) is strictly greater than its topological dimension (Euclidean dimension).

In the phenomenon of fracture, by monotonic loading test or impact on a piece of metal, ceramic, or polymer, as the chemical bonds between the atoms of the material are broken, it produces two complementary fracture surfaces. Due to the irregular crystalline arrangement of these materials the fracture surfaces can also be irregular, i.e., rough and difficult geometrical description. The roughness that they have is directly related to the material microstructure that are formed. Thus, the various microstructural features of a material (metal, ceramic, or polymer) which may be, particles, inclusions, precipitates, etc. affect the topography of the fracture surface, since the different types of defects present in a material can act as stress concentrators and influence the formation of fracture surface. These various microstructural defects interact with the crack tip, while it moves within the material, forming a totally irregular relief as chemical bonds are broken, allowing the microstructure to be separated from grains (transgranular and intergranular fracture) and microvoids are joining (coalescence of microvoids, etc..) until the fracture surfaces depart. Moreover, the characteristics of macrostructures such as the size and shape of the sample and notch from which the fracture is initiated, also influence the formation of the fracture surface, due to the type of test and the stress field applied to the specimen.

After the above considerations, one can say with certainty that the information in the fracture process are partly recorded in the "story" that describes the crack, as it walks inside the material [2]. The remainder of this information is lost to the external environment in a form of dissipated energy such as sound, heat, radiation, etc. [30, 31]. The remaining part of the information is undoubtedly related to the relief of the fracture surface that somehow describes the difficulty that the crack found to grow [2]. With this, you can analyze the fracture phenomenon through the relief described by the fracture surface and try to relate it to the magnitudes of fracture mechanics [3, 4, 5, 6, 7, 8, 9 - 11, 12, 13]. This was the basic idea that brought about the development of the topographic study of the fracture surface called fractography.

In fractography anterior the fractal theory the description of geometric structures found on a fracture surface was limited to regular polyhedra-connected to each other and randomly distributed throughout fracture surface, as a way of describing the topography of the irregular surface. Moreover, the study fractographic hitherto used only techniques and statistical analysis profilometric relief without considering the geometric auto-correlation of surfaces associated with the fractal exponents that characterize the roughness of the fracture surface.

The basic concepts of fractal theory developed by Mandelbrot [1] and other scientists, have been used in the description of irregular structures, such as fracture surfaces and crack [14], in order to relate the geometrical description of these objects with the materials properties [15].

The fractal theory, from the viewpoint of physical, involves the study of irregular structures which have the property of invariance by scale transformation, this property in which the parts of a structure are similar to the whole in successive ranges of view (magnification or reduction) in all directions or at least one direction (self-similarity or self-affinity, respectively) [36]. The nature of these intriguing properties in existing structures, which extend in several scales of magnification is the subject of much research in several phenomena in nature and in materials science [16, 17 and others]. Thus, the fractal theory has many contexts, both in physics and in mathematics such as chaos theory [18], the study of phase transitions and critical phenomena [19, 20, 21], study of particle agglomeration [22], etc.. The context that is more directly related to Fracture Mechanics, because of the physical nature of the process is with respect to fractal growth [23, 24, 25, 26]. In this subarea are studied the growth mechanisms of structures that arise in cases of instability, and dissipation of energy, such as crack [27, 28] and branching patterns [29]. In this sense, is to be sought to approach the problem of propagation of cracks.

The fractal theory becomes increasingly present in the description of phenomena that have a measurable disorder, called deterministic chaos [18, 27, 28]. The phenomenon of fracture and crack propagation, while being statistically shows that some rules or laws are obeyed, and every day become more clear or obvious, by understanding the properties of fractals [27, 28].

FUNDAMENTAL GEOMETRIC ELEMENTS AND MEASURE THEORY ON FRACTAL GEOMETRY

In this part will be presented the development of basic concepts of fractal geometry, analogous to Euclidean geometry for the basic elements such as points, lines, surfaces and fractals volumes. It will be introduce the measurement fractal theory as a generalization of Euclidean measure geometric theory. It will be also describe what are the main mathematical conditions to obtain a measure with fractal precision.

Analogy Between Euclidean and Fractal Geometry

It is possible to draw a parallel between Euclidean and fractal geometry showing some examples of self-similar fractals projected onto Euclidean dimensions and some self-affine fractals. For, just as in Euclidean geometry, one has the elements of geometric construction, in the fractal geometry. In the fractal geometry one can find similar objects to these

Euclidean elements. The different types of fractals that exist are outlined in Figure 1 to Figure 4.

FractaisBetween 0≤D≤1 (Similar To Point)
An example of a fractal immersed in Euclidean dimension I=d+1=1 with projection in d=0, similar to punctiform geometry, can be exemplified by the Figure 1.

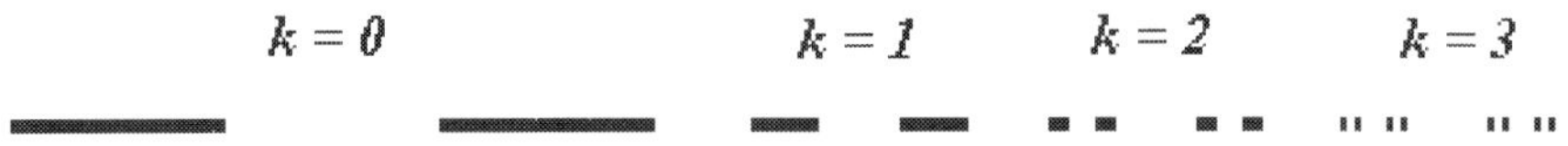

$$k = 0 \qquad\qquad k = 1 \qquad k = 2 \qquad k = 3$$

Figure 1. Fractal immersed in the one-dimensional space where D≅0,631.

This fractal has dimension D≅0,631. This is a fractal-type "stains on the floor." Other fractal of this type can be observed when a material is sprayed onto a surface. In this case the global dimension of the spots may be of some value between 0≤D≤1.

FractaisBetween 1≤D≤2 (Similar To Straight Lines)
For a fractal immersed in a Euclidean dimension I=d+1=2, with projectionin d=1, analogous to the linear geometry is a fractal-type peaks and valleys (Figure 2). Cracks may also be described from this figure as shown in Alves [37]. Graphs of noise, are also examples of linear fractal structures whose dimension is between 1≤D≤2.

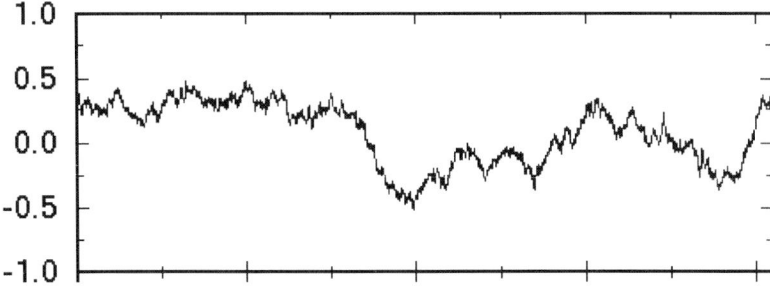

Figure 2. Fractal immersed in dimension d = 2. rugged fractal line.

Fractals Between 2≤D≤3 (Similar To Surfaces Or Porous Volumes)

For a fractal immersed in a Euclidean dimension, I=d+1=3 with projection in d=2, analogous to a surface geometry is fractal-type "mountains" or "rugged surfaces" (Figure 3). The fracture surfaces can be included in this class of fractals.

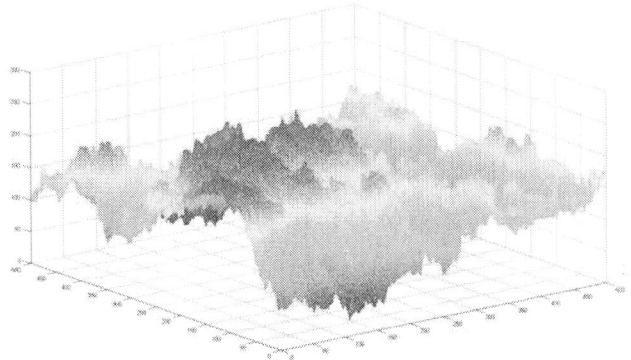

Figure 3. Irregular or rugged surface that has a fractal scaling with dimension D between 2≤D≤3.

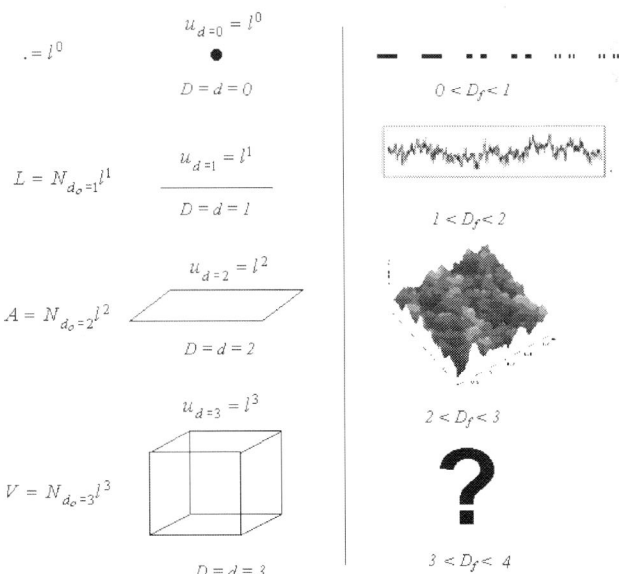

Figure 4. Comparison between Euclidean and fractal geometry. D,d and Df represents the topological, Euclidean and fractal dimensions, of a point, line segment, flat surface, and a cube, respectively

Making a parallel comparison of different situations that has been previously described, one has (Figure 4)

Fractal Dimension (Non-Integer)

An object has a fractal dimension, D, ($d \leq D \leq d+1=I$), where I is the space Euclidean dimension which is immersed, when:

$$F\left(\varepsilon L_0\right) = \varepsilon^{-D} F\left(L_0\right)$$

(1)

where L_0 is the projected length that characterizes an apparent linear extension of the fractal ε, is the scale transformation factor between two apparent linear extension, $F(L0)$ is a function of measurable physical properties such as length, surface area, roughness, volume, etc., which follow the scaling laws, with homogeneity exponent is not always integers, whose geometry that best describe, is closer to fractal geometry than Euclidean geometry. These functions depend on the dimensionality, I, of the space which the object is immersed. Therefore, for fractals the homogeneity degree n is the fractal dimension D (non-integer) of the object, where ε is an arbitrary scale.

Based on this definition of fractal dimension it can be calculates doing:

$$\varepsilon^{-D} = \frac{F(\varepsilon L_o)}{F(L_o)}$$

(2)

taking the logarithm one has

$$D = -\frac{\ln\left[\frac{F(\varepsilon L_o)}{F(L_o)}\right]}{\ln(\varepsilon)}$$

(3)

From the geometrical viewpoint, a fractal must be immersed into a integer Euclidean dimension, $I=d+1$. Its non-integer fractal dimension, D, it appears because the fill rule of the figure from the fractal seed which obeys some failure or excess rules, so that the complementary structure of the fractal seed formed by the voids of the figure, is also a fractal.

For a fractal the space fraction filled with points is also invariant by scale transformation, i.e.:

$$P(L_o) = \frac{F(\lambda L_o)}{F(L_o)} = \frac{1}{N(L_o)}$$

(4)

Thus,

$$\varepsilon^D = P(L_0) \text{ ou} N(L_0) = \varepsilon^{-D}$$

(5)

where $P(L_0)$ is a probability measure to find points within fractal object

Therefore, the fractal dimension can be calculated from the fllowing equation:

$$D = -\frac{\ln N(L_0)}{\ln \varepsilon}$$

(6)

If it is interesting to scale the holes of a fractal object (the complement of a fractal), it is observed that the fractal dimension of this new additional dimension corresponds to the Euclidean space in which it is immersed less the fractal dimension of the original.

A Generalized Monofractal Geometric Measure

Now will be described how to process a general geometric measure whose dimension is any. Similarly to the case of Euclidean measure the measurement process is generalized, using the concept of Hausdorff-Besicovitch dimension as follows.

Suppose a geometric object is recovered by α-dimensional, geometric units, uD, with extension, δk and $\delta k \leq \delta$, where δ is the maximum α-dimensional unit size and α is a positive real number. Defining the quantity:

$$M_D(\alpha, \delta, \{\delta_k\}) = \sum_k \delta_k{}^\alpha$$

(7)

Choosing from all the sets $\{\delta k\}$, that reduces this summation, such that:

$$M_D(\alpha, \delta) = \inf_{\{\delta_k\}} \sum_k \delta_k{}^\alpha$$

(8)

The smallest possible value of the summation in (8) is calculated to obtain the adjustment with best precision of the measurement performed. Finally taking the limit of δ tending to zero, (δ→0), one has:

$$M_D(\alpha) = \lim_{\delta \to 0} M_D(\alpha, \delta)$$

(9)

The interpretation for the function MD(α) is analogous to the function for a Euclidean measure of an object, i.e. it corresponds to the geometric extension (length, area, volume, etc.) of the set measured by units with dimension, α. The cases where the dimension is integer are same to the usual definition, and are easier to visualize. For example, the calculation of MD(α) for a surface of finite dimension, D=2, there are the cases:

- For α=1<D=2 measuring the "length" of a plan with small line segments, one gets MD=∞, because the plan has a infinity "length", or there is a infinity number of line segment inside the plane.
- For, α=2=D=2 measuring the surface area of small square, one gets MD=Ad=2=A0. Which is the only value of α where MD is not zero nor infinity (see Figure 5.)
- For α=3>D=2 measuring the "volume" of the plan with small cubes, one gets MD=0, because the "volume" of the plan is zero, or there is not any volume inside the plan.

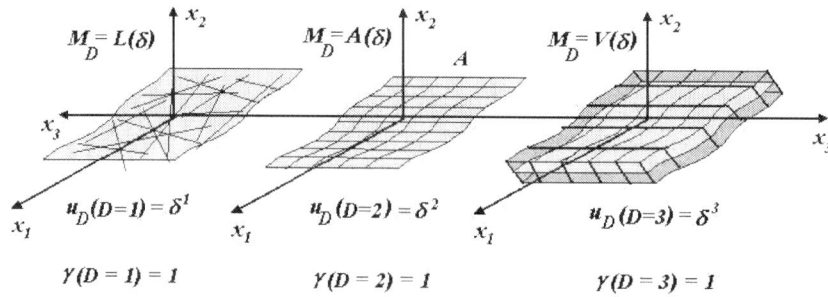

Figure 5. Measuring, MD(δ) of an area A with a dimension, D=2 made with different measure units u D for D=1,2,3.

Therefore, the function, MD possess the following form

$$M_D(\alpha) = \begin{cases} 0 \, para \, \alpha > D \\ M \, para \, \alpha = D \\ \infty \, para \, \alpha < D \end{cases}$$

$$(10)$$

That is, the function MD only possess a different value of 0 and ∞ at a critical point $\alpha=D$ defining a generalized measure

Invariance Condition of a Monofractal Geometric Measure

Therefore, for a generalized measurement there is a generalized dimension which the measurement unit converge to the determined value, M, of the measurement series, according to the extension of the measuring unit tends to zero, as shown in equations equações (9) and (10), namely:

$$M_D\left(\alpha, \delta, \{\delta_k\}\right) = \sum_k \delta_k{}^\alpha = M_{Do}\left(\delta\right) \varepsilon^{\alpha-D}$$

$$(11)$$

where $M_{D0}(\delta)$ is the Euclidean projected extension of the fractal object measured on α-dimensional space

Again the value of a fractal measure can be obtain as the result of a series. One may label each of the stages of construction of the function $M_D(\delta)$ as follows:

I. The first is the *measure itself*. Because it is actually the step that evaluates the extension of the set, summing the geometrical size of the recover units. Thus, the extension of the set is being overestimated, because it is always less or equal than tthe size of its coverage.

II. The next step is the *optimization* to select the arrangement of units which provide the smallest value measured previously, i.e. the value which best approximates the real extension of the assembly.

III. The last step is the *limit*. Repeat the previous steps with smaller and smaller units to take into account all the details, however small, the structure of the set.

As the value of the generalized dimension is defined as a critical function, $M\alpha=D(\delta)$ it can be concluded, wrongly, that the optimization step is not very important, because the fact of not having all its length measured accurately should not affect the value of critical point. The optimization step, this definition, serves to make the convergence to go faster in following step, that the mathematical point of view is a very desirable property when it comes to numerical calculation algorithms.

The Monofractal Measure and The Hausdorff-Besicovitch Dimension

In this part we will define the dimension-HausdorfBesicovicth and a fractal object itself. The basic properties of objects with "anomalous" dimensions (different from Euclidean) were observed and investigated at the beginning of this century, mainly by Hausdorff and Besicovitch [32,34]. The importance of fractals to physics and many other fields of knowledge has been pointed out by Mandelbrot [1]. He demonstrated the richness of fractal geometry, and also important results presented in his books on the subject [1, 35, 36].

The geometric sequence, S is given by:

$$S = \sum_k S_k \quad ondek = 0, 1, 2, \ldots.$$

(12)

represented in Euclidean space, is a fractal when the measure of its geometric extension, given by the series, $M\alpha(\delta k)$ satisfies the following Hausdorf-Besicovitch condition:

$$M_d(\delta_k) = \sum_k \gamma(d)\delta_k{}^\alpha = N_d(\delta_k)\gamma(d)\delta_k{}^\alpha \underset{\delta_k \to 0}{\left\langle \begin{array}{ll} 0; & \alpha > D \\ M_D; \alpha = D \\ \infty; & \alpha < D \end{array} \right.}$$

(13)

Where $\gamma(d)$ is the geometric factor of the unitary elements (or seed) of the sequence represented geometrically. δ: is the size of unit elements (or seed), used as a measure standard unit of the extent of the spatial representation of the geometric sequence. $N(\delta)$: is the number of elementary units (or seeds) that form the spatial representation of the sequence at a certain scaleα: the generalized dimension of unitary elements D: is the Hausdorff-Besicovitch dimension.

Fractal Mathematical Definition and Associated Dimensions

Therefore, fractal is any object that has a non-integer dimension that exceeds the topological dimension (D<I, where I is the dimension of Euclidean space which is immersed) with some invariance by scale transformation (self-similarity or self-affinity), where for any continuous contour that is taken as close as possible to the object, the number of points ND, forming the fractal not fills completely the space delimited by

the contour, i.e., there is always empty, or excess regions, and also there is always a figure with integer dimension, I, at which the fractal can be inscribed and that not exactly superimposed on fractal even in the limit of scale infinitesimal. Therefore, the fraction of points that fills the fractal regarding its Euclidean coverage is different of a integer. As seen in previous sections - 2.2 - 2.5 in algebraic language, a fractal is a invariant sequence by scale transformation that has a Hausdorff-Besicovitch dimension.

According to the previous section, it is said that an object is fractal, when the respective magnitudes characterizing features as perimeter, area or volume, are homogeneous functions with non-integer. In this case, the invariance property by scaling transformation (self-similar or self-affinity) is due to a scale transformation of at least one of these functions.

The fractal concept is closely associated to the concept of Hausdorff-Besicovitch dimension, so that one of the first definitions of fractal created by Mandelbrot [36] was:

"Fractal by definition is a set to which the Haussdorf-Besicovitch dimension exceeds strictly the topological dimension".

One can therefore say that fractals are geometrical objects that have structures in all scales of magnification, commonly with some similarity between them. They are objects whose usual definition of Euclidean dimension is incomplete, requiring a more suitable to their context as they have just seen. This is exactly the Hausdorff-Besicovitch dimension.

A dimension object, D, is always immersed in a space of minimal dimension $I=d+1$, which may present an excessive extension on the dimension d, or a lack of extension or failures in one dimension $d+1$. For example, for a crack which the fractal dimension is the dimension in the range of $1 \leq D \leq 2$ the immersion dimension is the dimension $I=2$ in the case of a fracture surface of which the fractal dimension is in the range $2 \leq D \leq 3$ the immersion dimension is the $I=3$. When an object has a geometric extension such as completely fill a Euclidean dimension regular, d, and still have an excess that partially fills a superior dimension $I=d+1$, in addition to the inferior dimension, one says that the object has a dimension in excess, de given by $de=D-d$ where D is the dimension of the object. For example, for a crack which the fractal dimension is in the range $1 \leq D \leq 2$ the excess dimension is $de=D-1$, in the case of a fracture surface of which the fractal dimension is in the range of $2 \leq D \leq 3$ the excess dimension is $de=D-2$. If on the other hand an object partially fills a Euclidean regular dimension, $I=d+1$ certainly this object fills fully a Euclidean regular dimension, d, so that it is said that this object has a lack dimension dfl=I−D=d+1−D, where de=1−dfl. For example, for a crack which the fractal dimension is the range of 1≤D≤2 the lack dimension is dfl=2−D. In the case of a fracture surface of which the fractal dimension is the range of 2≤D≤3 the lack dimension is dfl=3−D.

Classes and Types of Fractals

One of the most fascinating aspects of the fractals is the extremely rich variety of possible realizations of such geometric objects. This fact gives rise to the question of classification, and the book of Mandelbrot [1] and in the following publications many types of fractal structures have been described. Below some important classes will be discussed with some emphasis on their relevance to the phenomenon of growth.

Fractals are classified, or are divided into: mathematical and physical (or natural) fractals and uniform and non-uniform fractals. Mathematical fractals are those whose scaling relationship is exact, i.e., they are generated by exact iteration and purely geometrical rules and does not have cutoff scaling limits, not upper nor lower, because they are generated by rules with infinity interactions (Figure 6a) without taking into account none phenomenology itself, as shown in Figure 6a. Some fractals appear in a special way in the phase space of dynamical systems that are close to situations of chaotic motion according to the Theory of Nonlinear Dynamical Systems and Chaos Theory. This approach will not be made here, because it is another matter that is outside the scope of this chapter.

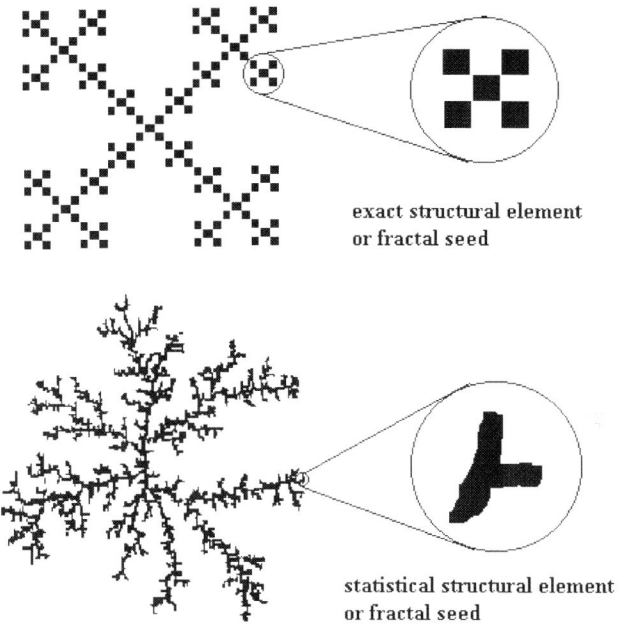

exact structural element
or fractal seed

statistical structural element
or fractal seed

Figure 6. Example of branching fractals, showing the structural elements, or elementary geometrical units, of two fractals. A) a self-similar mathematical fractal. B) a statistically self-similar physical fractal.

Real or physical fractals (also called natural fractals) are those statistical fracals, where not only the scale but all of fractal parameters can vary randomly. Therefore, their scaling relationship is approximated or statistical, i. e., they are observed in the statistical average made throughout the fractal, since a lower cutoff scale, ε min, to a different upper cutoff scale ε max (self-similar or self-affine fractals), as shown in Figure 6b. These fractals are those which appear in nature as a result of triggering of instabilities conditions in the natural processes [24] in any physical phenomenon, as shown Figure 6b. In these physical or natural fractals the extension scaling of the structure is made by means of a homogeneous function as follows:

$$F\left(\delta\right) \sim \delta^{d-D}$$

(14)

where d is the Euclidean dimension of projection of the fractal and D is the fractal dimension of self-similar structure.

It is true that the physical or real fractals can be deterministic or random. In random or statistical fractal the properties of self-similarity changes statistically from region to region of the fractal. The dimension cannot be unique, but characterized by a mean value, similarly to the analysis of mathematical fractals. The Figure 6b shows aspects of a statistically self-similar fractal whose appearance varies from branch to branch giving us the impression that each part is similar to the whole.

The mathematical fractals (or exact) and physical (or statistical), in turn, can be subdivided into uniform and nonuniform fractal.

Uniforms fractals are those that grow uniformly with a well behaved unique scale and constant factor, λ, and present a unique fractal dimension throughout its extension.

Non-uniform fractals are those that grow with scale factors λ_i's that vary from region to region of the fractal and have different fractal dimensions along its extension.

Thus, the fractal theory can be studied under three fundamental aspects of its origin:

1. From the geometric patterns with self-similar features in different objects found in nature.
2. From the nonlinear dynamics theory in the phase space of complex systems.
3. From the geometric interpretation of the theory of critical exponents of statistical mechanics.

METHODS FOR MEASURING LENGTH, AREA, VOLUME AND FRACTAL DIMENSION

In this section one intends to describe the main methods for measuring the fractal dimension of a structure, such as: the compass method, the Box-Counting method, the Sand-Box Method, etc.

It will be described, from now, how to obtain a measure of length, area or fractal volume. In fractal analysis of an object or structure different types of fractal dimension are obtained, all related to the type of phenomenon that has fractality and the measurement method used in obtaining the fractal measurement. These fractal dimensions can be defined as follows.

The Different Fractal Dimensions and Its Definitions

A fractal dimension Df in general is defined as being the dimension of the resulting measure of an object or structure, that has irregularities that are repeated in different scales (a invariance by scale transformation). Their values are usually noninteger and situated between two consecutive Euclidean dimensions called projection dimension d of the object and immersion dimension, d+1, i.e. $d \leq Df \leq d+1$.

In the literature there is controversy concerning the relationship between different fractal dimensions and roughness exponents. The term "fractal dimension" is used generically to refer to different fractional dimensions found in different phenomenologies, which results in formation of geometric patterns or energy dissipation, which are commonly called fractals [1]. Among these patterns is the growth of aggregates by diffusion (DLA - Diffusion Limited Aggregation), the film growth by ballistic deposition (BD), the fracture surfaces (SF), etc. The fractal dimensions found in these phenomena are certainly not the same and depend on both the phenomenology studied as the fractal characterization method used. Therefore, to characterize such phenomena using fractal geometry, a distinction between the different dimensions found is necessary.

Among the various fractal dimensions one can emphasize the Hausdorff-Besicovitch dimension, DHB, which comes from the general mathematical definition of a fractal [32, 33,34]. Other dimensions are the dimension box, DB, the roughness dimension or exponent Hurst, H, the Lipshitz-Hölder dimension, α, etc.. Therefore, a mathematical relationship between them needs to be clearly established for each phenomenon involved. However, is observed, then that relationship is not unique and depends not only on phenomenology, but also the characterization method used.

Therefore, the phenomenological equation of the fracture phenomenon can also, in theory, provide a relationship between fractal dimension and roughness exponent of a fracture surface, as happens to other phenomenologies. In this study, there was obtained a fractal model for a fracture surface, as a generalization of the box-counting method. Thus, will be discussed the relationship between the local and global box dimension and the roughness dimension, which are involved in the characterization of a fracture surface, and any other dimension necessary to describe a fractal fracture surface.

Compass Methods and Divider Dimension, D_D

The divider dimension D_D is defined from the measure of length of a roughened fractal line, for example, when using the compass method. This measure is obtained by opening a compass with an aperture δ and moving on the line fractal to obtain the value of the line length rugosa (see Figure 7). The different values of the rough line length due to the compass aperture determines the dimension divider.

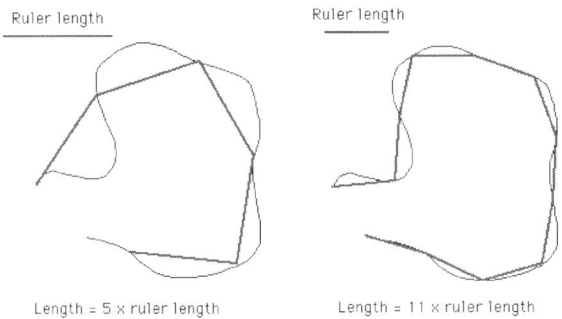

Figure 7. Compass method applied to a rugged line.

For a fractal rough line the divider dimension can be defined as:

$$D_D \equiv -\frac{\ln\left(\frac{L}{\delta}\right)}{\ln\left(\frac{\delta}{L_0}\right)}$$

(15)

where L_0 is the projected length obtained from the rugged fractal length L

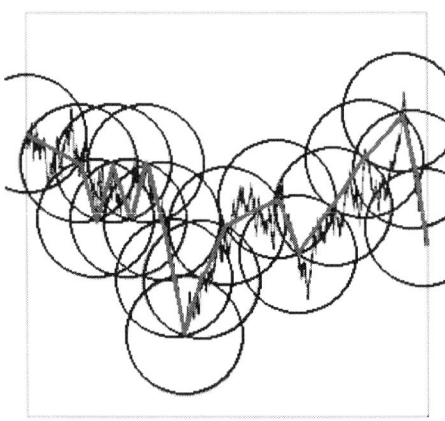

Figure 8. Compass method applied on a line noise or a rough self-affine fractal.

Several methods for determining the fractal dimension based on the compass method, among them stand out the following methods: the Coastlines Richardson Method, the Slit Island Method, etc.

Methods of Measurement for Determining the Fractal Dimension of a Structure

There are basically two ways to recover an object with boxes for fractal dimension measuring. In the first method, boxes of different sizes extending from a minimum size δ min until to a maximum size δ max, from a fixed origin recovering the whole object at once time. In the second case, one side of the recovering box is kept fixed, and with a minimum size ruler, δ min, then recovers the figure by moving the boundary of that recovering from the minimum δ min to maximum size δ max of the object. The first method is known as a method Box-Counting exemplified in Figure 9 and the second method is known as Sand-box, shown in Figure 10. The advantage of the second over the first is that it detects the changes in dimension D with the length of the object. If the object under consideration has a local dimension for boxes with size $\delta \rightarrow 0$, unlike the global dimension, $\delta \rightarrow \infty$, it is said that the object is self-affine fractal. Otherwise the object is said self-similar. These two main methods of counts of structures which may lead to determination of the fractal dimension of an object [38].

Box-Counting Method by Static Scaling of the Elements in a Fractal Structure

The Box-Counting method, comes from the theory of critical phenomena in statistical mechanics. In statistical mechanics there is an analogous mathematical method to describing phenomena which have self-similar properties, permitting scale transformations without loss of generality in the description of physical information of the phenomenon ranging from quantities such as volume up to energy. However, in the case described here, the Box-Counting method is performed filling the space occupied by a fractal object with boxes of arbitrary size δ, and count the number $N(\delta)$ of these boxes in function its size, (Figure 9 and Figure 10). This number $N(\delta)$ of boxes is given as follows:

$$N(\delta)=C\delta^{D} \qquad (16)$$

Plotting the data in a log×log graph one obtains from the slope of the curve obtained, the fractal dimension of the object.

In the Box-Counting method (Figure 9), a grid that recover the object is divided into nk=L_0/δk boxes of equal side δk and how many of these boxes that recovering the object is counted. Then, varies the size of the boxes and the counting is retraced, and so on. Making a logarithm graph of the number Nk of boxes that recovering the object in function of the scale for each subdivision (εk=δk/L_0), one obtains the fractal dimension from the slope of this plot. Note that in this case the partition maximum is reached when, N∞=L0/δk(k→∞)=L_0/l0, where L_{max}=L0is the projected crack length δ∞=l0 is the length of the shortest practicable ruler.

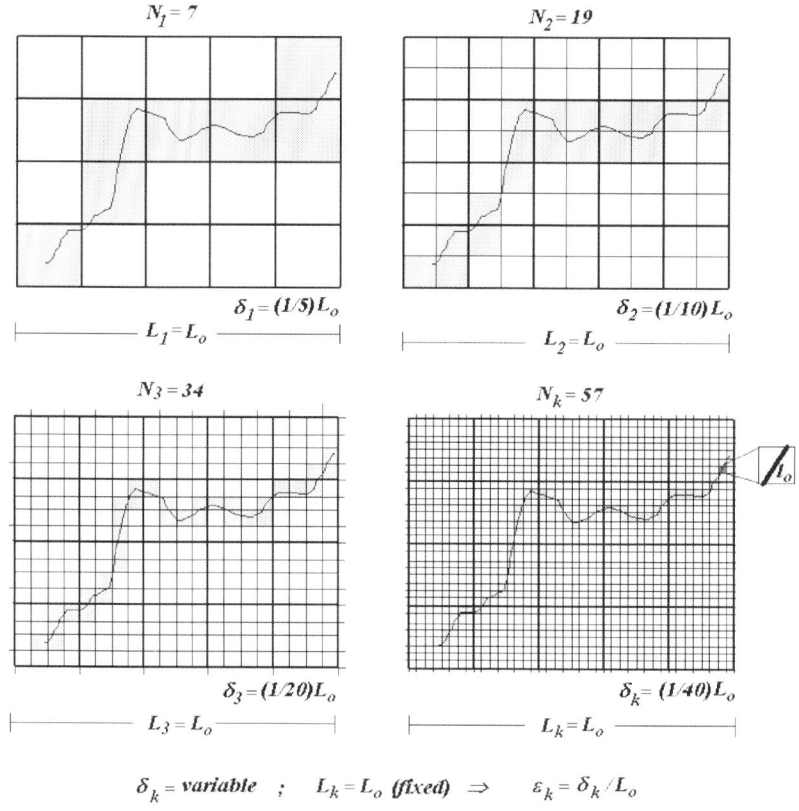

Figure 9. Fragment of a crack on a testing sample showing the variation of measurement of the crack length L with the measuring scale, εk=δk/L0 for a partition, δk=variável and Lk=L0 (fixed), with sectioning done for counting by one-dimensional Box-Counting scaling method.

Therefore, the number Nk(δk) depending on the size, δk, of these boxes is given as follows:

$$N_k\left(\delta_k\right) = \left(\frac{\delta_k}{\delta_{max}}\right)^{-D}$$

(17)

In the Figure 9 is illustrated the use of this method in a fractal object. Are present different grids, or meshes, constructed to recover the entire structure, whose fractal dimension one wants to know. The grids are drawn from an original square, involving the whole space occupied by the structure. At each stage of refinement of the grid (L_0) (the number of equal parts in the side of the square is divided) are counted the number of squares $N(L_0)$ which contain part of the structure. Repeatedly from the data found, is constructed the graph of $\log L_0 \times \log N(L_0)$. If the graph thus obtained is a straight line, then the fractal behavior of the structure has self-similarity or statistical self-affinity whose dimension D is obtained by calculating the slope of the line. For more compact structure, it is recommended to make a statistical sampling, that is, the repeat the counting of the squares $N(L_0)$ for different squares constructed from the gravity center (counting center) of the in the structure. Thus, one obtains a set of values $N(L_0)$ for another set of values L_0. These data must be statistically treated to obtain the value of fractal dimension, "D".

From the viewpoint of experimental measurement, one can consider using different methods of viewing the crack to obtain the fractal dimension, such as optical microscopy, electron microscopy, atomic force microscope, etc.., Which naturally have different rules δk and therefore different scales of measurement εk,.

The fractal dimension is usually calculated using the Box-Counting shown in Figure 9, i.e. by varying the size of the measuring ruler δk and counting the number of boxes, Nk that recover the structure. In the case of a crack the fractal dimension is obtained by the following relationship:

$$D = -\frac{\ln N}{\ln(l_o/L_o)}$$

(18)

The description of a crack according to the Box-Counting method follows the idea shown in Figure 9, which results in:

$$D = -\frac{\ln 57}{\ln(1/40)} = 1.096$$

(19)

The same result can be obtained using the Box-Sand method, as shown in Figure 10.

The Sand-Box Counting Method of the Elements by Static Scaling of a Fractal Structure

The Sand-box method consists in the same way as the Box-Counting method, to count the number of boxes, N(u), but with fixed length, u, as small as possible, extending gradually up the boundary count until to reach out to the border of the object under consideration. This is done initially by setting the counting origin from a fixed point on the object, as shown in Figure10. This method seems to be the most advantageous, as well as to establish a coordinate system, or a origin for calculating the fractal dimension, it also allows, in certain cases, to infer dynamic data from static scaling, as shown by Alves [47].

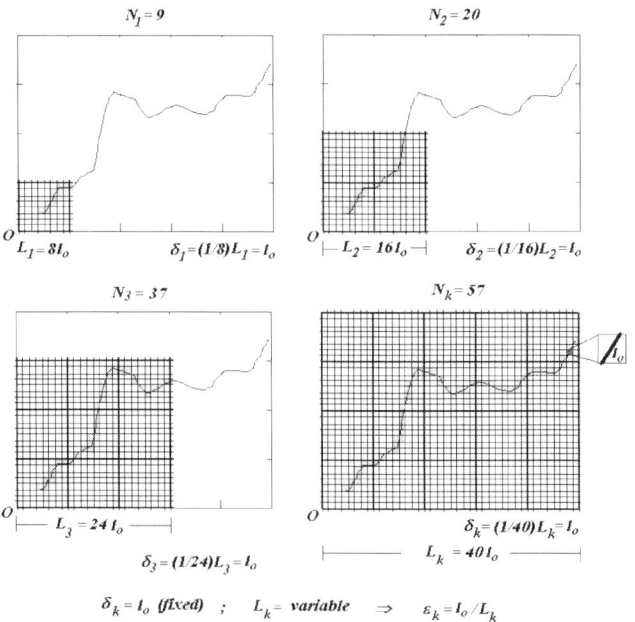

Figure 10. Fragment of a crack on test specimen showing the variation of measurement of the crack length L with the measuring scale, $\varepsilon_k = \delta_k/L_0$ for a partition Lk=variável, and δ_k=l0 (fixed), with sectioning done for counting by one-dimensional Sand-Box scaling method.

In the Sand-Box method (Figure10), the figure is recovered with boxes of different sizes Lk, no matter the form, which can be rectangular or spherical, however, fixed at a any point "O" on figure called origin, from which the boxes are enlarged. It is counted the number of elementary structures, or seeds, which fit within each box. Plotting the graph of log $N_k \times \log (\varepsilon_k = \delta_{min}/L_k)$ in the same manner as in the above method the fractal dimension is obtained. Note that in this case the maximum partition is achieved when $N\infty = L_k(k\to\infty)/\delta_{min} = L_0/l0$, where $L_\infty = L_0$ is the projected crack length and $\delta_{min} = l0$ it is the length of the lower measuring ruler practicable.

The Global and Local Box Dimensions

To define the box dimension, DB, is assumed that all the space containing the fractal is recovered with a grid (set of α-dimensional units juxtaposed in the same shape and size, δ) with maximum size, δmax, which inscribes the fractal object. Defining the relative scale, ε on the grid size, δmax, as being given by:

$$\varepsilon = \frac{\delta}{\delta_{max}}$$

(20)

countting the number of boxes $N(\varepsilon)$ that have at least one point of the fractal. The box dimension is therefore defined as:

$$D_B = -\lim_{\varepsilon \to 0} \frac{\ln N(\varepsilon)}{\ln \varepsilon}$$

(21)

At this point, there are two ways to obtain the actual value of the measure, or taking the limit when $\varepsilon \to 0$ and allows that the dimension D fits the end value of $N(\varepsilon)$, or it is considered a linear correlation in value of $\ln N(\varepsilon) \times \ln \varepsilon$, which D is the slope of the line, and this defines the measure independently of the scale.

In the case of numerical estimation, one can not solve the limit indicated in the equation (21). Then, DB is obtained as a slope, $\ln N(\varepsilon) \times \ln \varepsilon$, when ε it is small. The value $N(\varepsilon)$ is obtained by an algorithm known as *Box-Counting*.

Self-affine fractals requiring different variations in scale length for different directions. Therefore, one can use the Box-Counting method with some care being taken, in the sense that the box dimension DB to be obtained has a crossing region between a local and global measure of the dimensions. From which follows that for each region is used the following relationships:

$$\lim_{l_0 \to 0} N(L_0) = \left(\frac{L_0}{l_0}\right)^{D_{Bg}} p/L_0 << L_{0s}$$

(22)

for a global measurement

$$\lim_{l_0 \to 0} N(L_0) = \left(\frac{L_0}{l_0}\right)^{D_{Bl}} p/L_0 >> L_{0s}$$

(23)

where L_{0s} is the threshold saturation length which the fractal dimension changes its behavior from local to global stage.

For measurement, generally, for any self-affine fractal structure the local fractal dimension is related to the Hurst exponent, H, as follow,

$$D_{Bl} = d + 1 - H_{q=1}$$

(24)

At this point, one observes that for a profile the relationship DB l=2−H commonly used, only serves for a local measurements using the box counting method. While for global measures one can not establish a relationship between DBg and H. For the global fractal dimension, Dg=d and l=d+1 the Euclidean dimension where the fractal is embedded one has

$$d \leq D_{Bg} \leq d + 1$$

(25)

Some textbooks on the subject show an example of calculation of local and global fractal dimension of self-affine fractals, obtained by a specific algorithm [18, 22, 23, 26, 38,39].

In crossing the limit of fractal dimension local Dl to global Dg, there is a transition zone called the "crossover", and the results obtained in this region are somewhat ambiguous and difficult to interpret [39]. However, in the global fractal dimension, the structure is not considered a fractal [42, 43].

The Relationship between Box Dimension and Hausdorff besicovitch Dimensions

The mathematical definition of generalized dimension of Haussdorff-Besicovitch need a method that can measure it properly to the fractal phenomenon under study. Some authors [23, 40, 44, 45] have discussed the possibility of using the Box-Counting method as one of the graphical methods which obtains a box dimension DB, very close to generalized HaussdorffBesicovitch, DHB, i.e. [44]:

$$D_B \cong D_{HB} \qquad (26)$$

In this sense the box dimension, DB is obtained for self-asimilar fractals that may be rescaled for the same variation in scales lengths in all directions by using the relationship:

$$N(L_0) = \left(\frac{L_0}{l_0}\right)^{D_B} \qquad (27)$$

where l_0 is the grid size used and L0 is the apparent size of the fractal to be characterized.

The analytical calculation of the Hausdorff dimension is only possible in some cases and it is difficult to implement by computation. In numerical calculation, is used another more appropriate definition, called box dimension, DB, which in the case of dynamic systems, has the same value of the Haussdorff dimension, D [44]. Thus, it is common to call them without distinction as *fractal dimensions*, D as will be shown below.

All the definitions related to fractal exponents that are shown here, and all numerical evaluation of these, always calculates the inclination of some amount ε against on a logarithmic scale.

The two definitions of, *Hausdorff-Besicovitch Dimension*, DH and *Box-Dimension*, DB are allocated the same amount, but in a way somewhat different from each other. In inaccurate way, one can think that the connection between the two is done considering that:

$$M_D(\alpha \to D) \sim N(\varepsilon)\varepsilon^d, \qquad (28)$$

by analogy with equation (13), i.e. approximating to the geometric extension of the object by the number of boxes (of the same size) necessary to recover it. But, since the definition of the box dimension there is no optimization step, and its value is directly dependent on $N(\varepsilon)$ (which is not the case with the Hausdorff dimension) in practice one has often the geometric extension is overestimated, particularly for ε large, i. e. upper limit $(\varepsilon \rightarrow 1)$ and thus DB$\leq$D. However, for the lower limit, i.e. $\varepsilon \rightarrow 0$, the Hausdorff-Besicovitch dimensions, DH and the box dimension, DB are equal, becoming valid the measure of geometric extension process, $MD(\delta)$ at box counting algorithm.

Considering from (28) that:

$$N(\varepsilon) \sim \varepsilon^{-D} (d \leq D \leq d+1) \tag{29}$$

and that

$$N(\varepsilon_{max}) \sim \varepsilon_{max}^{-D} (d \leq D \leq d+1) \tag{30}$$

Therefore, dividing (29) by (30) has:

$$N(\varepsilon)N(\varepsilon_{max}) \sim (\varepsilon_{max})^{-D} (d \leq D \leq d+1) \tag{31}$$

taking ε_{max} the total grid extension that recover the object, one has:

$$\varepsilon_{max} \rightarrow 1 \tag{32}$$

From as early as (31)

$$N(\varepsilon) \rightarrow \varepsilon^{-D} (d \leq D \leq d+1) \tag{33}$$

Substituting (33) in (28) has:

$$M_D(\alpha \rightarrow D) \sim \varepsilon^{\alpha-D}, \tag{34}$$

This equation is analogous to the fundamental Richardson relationship for a fractal length.

CRACK AND RUGGED FRACTURE SURFACE MODELS

The two main problematics of mathematical description of Fracture Mechanics are based on the following aspects: the surface roughness generated in the process and the field stress/strain applied to the specimen. This section deals with the fractal mathematical description of the first aspect, i.e., the roughness of cracks on Fracture Mechanics, using fractal geometry to model its irregular profile. In it will be shown basic mathematical assumptions to model and describe the geometric structures of irregular cracks and generic fracture surfaces using the fractal geometry. Subsequently, one presents also the proposal for a self-affine fractal model for rugged surfaces of fracture. The model was derived from a generalization of Voss [48] ([1] -) equation and the model of Morel [49] for fractal self-affine fracture surfaces. A general analytical expression for a rugged crack length as a function of the projected length and fractal dimension is obtained. It is also derived the expression of roughness, which can be directly inserted in the analytical context of Classical Fracture Mechanics.

The objectives of this section are: (i) based geometrical concepts, extracted from the fractal theory and apply them to the CFM in order to (ii) construct a precise language for its mathematical description of the CFM, into the new vision the fractal theory. (iii) eliminate some of the questions that arise when using the fractal scaling in the formulation of physical quantities that depend on the rough area of fracture, instead of the projected area, in the manner which is commonly used in fracture mechanics. (iv) another objective is to study the way which the fractal concept can enrich and clarify various aspects of fracture mechanics. For this will be done initially in this section, a brief review of the major advances obtained by the fractal theory, in the understanding of the fractography and in the formation of fracture surfaces and their properties. Then it will be done, also, a mathematical description of our approach, aiming to unify and clarify aspects still disconnected from the classical theory and modern vision, provided by fractal geometry. This will make it possible for the reader to understand what were the major conceptual changes introduced in this work, as well as the point from which the models proposed progressed unfolding in new concepts, new equations and new interpretations of the phenomenon.

Application of Fractal Theory in the Characterization of a Fracture Surface

In this section one intends to do a brief history of the fractography development as a fractal characterization methodology of a fractal fracture surface.

Geometric Aspects and Observations Extracted from the Quantitative Fractography of Irregular Fracture Surface

The technique used for geometric analysis of the fracture surface is called fractography. Until recently it was based only on profilometric study and statistical analysis of irregular surfaces [50]. Over the years, after repeated observations of these surfaces at various magnifications, was also revealed a variety of self-similar structures that lie between the micro and macrostructural level, characteristic of the type of fracture under observation. Since 1950 it is known that certain structures observed in fracture surfaces by microscopy, showed the phenomenon of invariance by magnification. Such structures recently started to be described in a systematic way by means of fractal geometry [51, 25]. This new approach allows the description of patterns that at first sight seem irregular, but keep an invariance by scale transformation (self-similarity or self-affinity). This means that some facts concerning the fracture have the same character independently of the magnification scale, i.e. the phenomenology that give rise to these structures is the same in different observation scales.

The Euclidean scaling of physical quantities is a common occurrence in many physical theories, but when it comes to fractality appears the possibility to describe irregular structures. The fracture for each type of material has a behavior that depends on their physical, chemical, structural, etc. properties. Looking at the topography and the different structures and geometrical patterns formed on the fracture surfaces of various materials, it is impossible to find a single pattern that can describe all these surfaces (Figure 11), since the fractal behavior of the fracture depends on the type of material [52]. However, the fracture surfaces obtained under the same mechanical testing conditions and for the same type of material, retains geometric aspects similar of its relief [53] (see Figure 12).

This similarity demonstrates that exist similar conditions in the fracture process for the same material, although also exist statistically changinga from piece to piece, constructed of the same material and under the same conditions [54; 55]. Based on this observation was born the idea to relate the surface roughness of the fracture with the mechanical properties of materials [50].

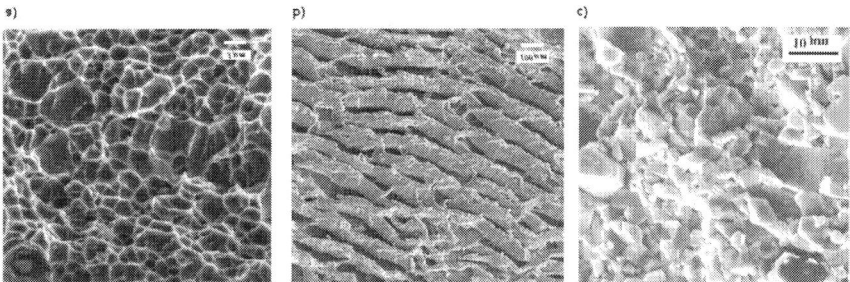

Figure 11. Various aspects of the fracture surface for different materials: (a) Metallic B2CT2 sample, (b) Polymeric, sample PU1.0, with details of the microvoids formation during stable crack propagation, (c) Ceramic [56].

Fractal Theory Applied to Description of The Relief of a Fracture Surface

Let us now identify the fractal aspects of fracture surfaces of materials in general, to be obtained an experimental basis for the fractal modeling of a generic fracture surface. The description of irregular patterns and structures, is not a trivial task. Every description is related to the identification of facts, aspects and features that may be included in a class of phenomena or structure previously established. Likewise, the mathematical description of the fracture surface must also have criteria for identifying the geometric aspects, in order to identify the irregular patterns and structures which may be subject to classification. The criteria, used until recently were provided by the fractográfico study through statistical analysis of quantities such as average grain size, roughness, etc. From geometrical view point this description of the irregular fracture surface, was based, until recently, the foundations of Euclidean geometry. However, this procedure made this description a task too complicated. With the advent of fractal geometry, it became possible to approach the problem analytically, and in more authentic way.

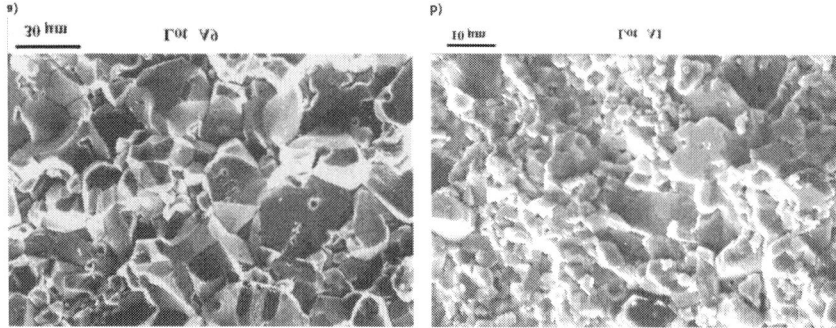

Figure 12. Fracture surfaces of different parts mades with the same material, a) Lot A9 b) Lot A1 [56 1999].

Inside the fractography, fractal description of rugged surfaces, has emerged as a powerful tool able to describe the fracture patterns found in brittle and ductile materials. With this new characterization has become possible to complement the vision of the fracture phenomenon, summarizing the main geometric information left on the fracture surface in just a number, "D", called fractal dimension. Therefore, assuming that there is a close relationship between the physical phenomena and fractal pattern generated as a fracture surface, for example, the physical properties of these objects have implications on their geometrical properties. Thinking about it, one can take advantage of the geometric description of fractals to extract information about the phenomenology that generated it, thereby obtaining a greater understanding of the fracture process and its physical properties. But before modeling any irregular (or rough) fracture surface, using fractal geometry, will be shown some of the difficulties existing and care should be taken in this mathematical description.

Fractal Models of a Rugged Fracture Surface

A fracture surface is a record of information left by the fracture process. But the Classical Fracture Mechanics (CFM) was developed idealizing a regular fracture surface as being smooth and flat. Thus the mathematical foundations of CFM consider an energy equivalence between the rough (actual) and projected (idealized) fracture surfaces [57]. Besides the mathematical complexity, part of this foundation is associated with the difficulties of an accurate measure of the actual area of fracture. In fact, the geometry of the crack surfaces is usually rough and can not be described in a mathematically simple by Euclidean geometry [52]. Although there are several methods to quantify the fracture area, the results are dependent on the measure ruler size used [56]. Since the last century all the existing methods to measure a rugged surface did not contribute to its insertion into the analytical mathematical formalism of CFM until to rise the fractal geometry. Generally, the roughness of a fracture surface has fractal geometry. Therefore, it is possible to establish a relationship between its topology and the physical quantities of fracture mechanics using fractal characterization techniques. Thus, with the advent of fractal theory, it became possible to describe and quantify any structure apparently irregular in nature [1]. In fact, many theories based on Euclidean geometry are being revised. It was experimentally proved that the fracture surfaces have a fractal scaling, so the Fracture Mechanics is one of the areas included in this scientific context.

Importance of Fracture Surface Modeling

The mathematical formalism of the CFM was prepared by imagining a fracture surface flat, smooth and regular. However, this is an mathematical idealization because actually the microscopic viewpoint, and in some cases up to macroscopic a fracture surface is generally a rough and irregular structure difficult to describe geometrically. This type of mathematical simplification above mentioned, exists in many other areas of exact sciences. However, to make useful the mathematical formalism developed over the years, Irwin started to consider the projected area of the fracture surface [57] as being energetically equivalent to the rugged surface area. This was adopted due to experimental difficulties to accurately measure the true area of the fracture, in addition to its highly complex mathematics. Although there are different methods to quantify the actual area of the fracture [56], its equationing within the fracture mechanics was not considered, because the values resulting from experimental measurements depended on the "ruler size" used by various methods. No mathematical theory had emerged so far, able to solve the problem until a few decades came to fractal geometry. Thus, modern fractal geometry can circumvent the problem of complicated mathematical description of the fracture surface, making it useful in mathematical modeling of the fracture.

In particular, it was shown experimentally that cracks and fracture surfaces follow a fractional scaling as expected by fractal geometry. Therefore, the fractal modeling of a irregular fracture surface is necessary to obtain the correct measurement of its true area. Therefore, fracture mechanics is included in the above context and all its classical theory takes into account only the projected surface. But with the advent of fractal geometry, is also necessary to revise it by modifying its equations, so that their mathematical description becomes more authentic and accurate. Thus, it is possible to relate the fractal geometric characterization with the physical quantities that describe the fracture, including the true area of irregular fracture surface instead of the projected surface. Thought this idea was that Mandelbrot and Passoja [58] developed the fractal analysis by the "slit island method ". Through this method, they sought to correlate the fractal dimension with the physical well-known quantities in fracture mechanics, only an empirical way. Following this pioneering work, other authors [3, 4, 5, 6, 7, 8, 11,12, 13, 59] have made theoretical and geometrical considerations with the goal of trying to relate the geometrical parameters of the fracture surfaces with the magnitudes of fracture mechanics, such as fracture energy, surface energy, fracture toughness, etc.. However, some misconceptions were made regarding the application of fractal geometry in fracture mechanics.

Several authors have suggested different models for the fracture surfaces [60-63]. Everyone knows that when it was possible to model generically a fracture surface, independently of the fractured material, this will allow an analytical description of the phenomena resulting the roughness of these surfaces within the Fracture Mechanics. Thus the Fracture Mechanics will may incorporate fractal aspects of the fracture surfaces explaining more appropriately the material properties in general. In this section one propose a generic model, which results in different cases of fracture surfaces, seeking to portray the variety of geometric features found on these surfaces for different materials. For this a basic mathematical conceptualization is needed which will be described below. For this reason it is done in the following section a brief bibliographic review of the progress made by researchers of the fractal theory and of the Fracture Mechanics in order to obtain a mathematical description of a fracture surface sufficiently complete to be included in the analytical framework of the Mechanics Fracture.

Literature Review - Models of Fractal Scaling of Fracture Surfaces

Mosolov [64] and Borodich [3] were first to associate the deformation energy and fracture surface involved in the fracture with the exponents of surface roughness generated during the process of breaking chemical bonds, separation of the surfaces and consequently the energy dissipation. They did this relationship using the stress field. Mosolov and Borodich [64, 3] used the fractional dependence of singularity exponents of this field at the crack tip and the fractional dependence of fractal scaling exponents of fracture surfaces, postulating the equivalence between the variations in deformation and surface energy. Bouchaud [62] disagreed with the Mosolov model [64] and proposed another model in terms of fluctuations in heights of the roughness on fracture surfaces in the perpendicular direction to the line of crack growth, obtaining a relationship between the fracture critical parameters such as KIC and relative variation of the height fluctuations of the rugged surface. In this scenario has been conjectured the universality of the roughness exponent of fracture surfaces because this did not depend on the material being studied [63]. This assumption has generated controversy [61] which led scientists to discover anomalies in the scaling exponents between local and global scales in fracture surfaces of brittle materials. Family and Vicsék [39, 65] and Barabasi [66] present models of fractal scaling for rugged surfaces in films formed by ballistic deposition. Based on this dynamic scaling Lopez and Schimittibuhl [67, 68] proposed an analogous model valid for fracture surfaces, where they observed in your experiments anomalies in the fractal scaling, with critical dimensions of transition for the behavior of the roughness of these surfaces in brittle materials. In this sense Lopez [67, 68] borrowed from the model of Family and Vicsék [39, 65] analogies that could be applied to the rough fracture surfaces.

The Fractality of a Crack or Fracture Surface

By observing a crack, in general, one notes that it presents similar geometrical aspects that reproduce itself, at least within a limited range of scales. This property called invariance by scale transformation is called also self-similarity, if not privilege any direction, or self-affinity, when it favors some direction over the other. Some authors define it as the property that have certain geometrical objects, in which its parts are similar to the whole in in successive scales transformation. In the case of fracture, this takes place from a range of minimum cutoff scale,εmin until a maximum cutoff scale, εmax, contrary to the proposed by Borodich [3], which defines an infinite range of scales to maintain the mathematical definition fractal. In the model proposed in this section, one used the fractal theory as a form closer to reality to describe the fracture surface with respect to Euclidean description. This was done in order to have a much better approximatation to reality of the problem and to use fractal theory as a more authentic approach.

To understand clearly the statements of the preceding paragraph, one can use the pine example shown in Figure 13. It is known that any stick of a pine is similar in scale, the other branches, which in its turn are similar to the whole pine. The relationship between the scales mentioned above, in case of pine, can be obtained considering from the size of the lower branch (similar to the pine whole) until the macroscopic pine size. Calling of $\delta min=l0$, the size of the lower branch and $\delta max=L0$, the macroscopic size of whole pine one may be defined cutoff scales lower and upper (minimum and maximum), subdivided, therefore, the pine in discrete levels of scales as suggested the structure, as follows:

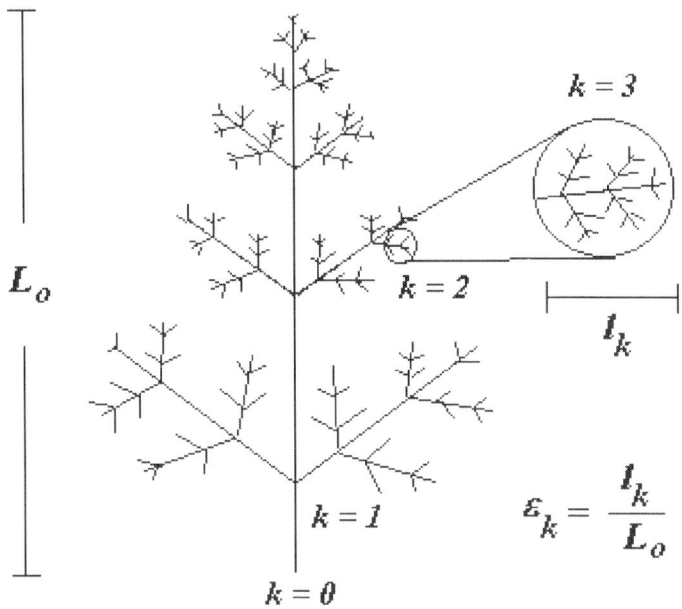

Figure 13. Self-similarity present in a pine (fractal), with different levels of scaling, k.

$$\varepsilon_{min} = \frac{l_o}{L_o} \leq \varepsilon_k = \frac{l_k}{L_o} \leq \varepsilon_{max} = \frac{L_o}{L_o} = 1 \begin{cases} static\,case\,L_0 = L_{0\,max} \\ dynamic\,case\,L_0 = L_0(t) \end{cases}$$

(35)

where an intermediate scale $\varepsilon_k(\varepsilon_{min}\leq\varepsilon k\leq\varepsilon_{max})$ can also be defined as follows:

$$\varepsilon_k = \frac{l_k}{L_o}.$$

(36)

The magnitude εk represents the scaling ratio which depicts the size of any branch with length, lk, in relation to any pine whole. l0 is related to the Mishnaevsky minimum size for a crack which is shown in section - 4.2.6 and L0=L0max is the maximum leght if the fracture already been completed.

Similarly it is assumed that the cracks and fracture surfaces also have their scaling relations, like that represented in equations (35) and (36). In the continuous cutoff scale levels, lower and upper (minimum and maximum), are thus defined as follows:

$$\varepsilon_{min} = \frac{l_o}{L_o} \leq \varepsilon = \frac{l}{L_o} \leq \varepsilon_{max} = \frac{L_o}{L_o} = 1$$

(37)

Note that the self-similarity of the pine so as the crack self-affinity, although statistical, is limited by a lower scale ε_{min} as determined by the minimum size, l0, and a upper scale ε_{max}, given by macroscopic crack size, L0.

From the concepts described so far, it is verified that the measuring scale εk to count the structure elements is arbitrary. However, in the scaling of a fracture surface, or a crack profile, follows a question:

Which is the value of scale ε_k to be properly used in order to obtain the most accurate possible measurement of the rugged fracture surface?

There is a minimum fracture size that depends only on the type of material?

Surely the answer to this question lies in the need to define the smallest size of the fractal structure of a crack or fracture surface, so that its size can be used as a minimal calibration measuring ruler ([2] -).

Since an fracture surface or crack, is considered a fractal, first, it is necessary to identify in the microstructure of the material which should be the size as small as possible a of a rugged fracture, i.e. the value of l_{min}. This minimal fracture size, typical of each material, must be then regarded as an elementary structure of the formation of fractal fracture, so defining a minimum cutoff scale, ε_{min}, for the fractal scaling, where $\varepsilon_{min}=l0/L0$, where l_0 it is a planar projection of l_{min}. In practice, from this value the minimum scale of measurement, $\varepsilon_{min}=l_0/L_0$ one defines a minimum ruler size δmin, for this case, equal to the value of the plane projection the smallest possible fracture size, i.e. $\delta min=l0$. Thus, the fractal scaling of the fracture surface, or crack, may be done by obtaining the most accurate possible value of the rough length, L. However, the theoretical prediction of the minimum fracture, $lmin$, must be made from the classical fracture mechanics, as will be seen below.

Scaling Hierarchical Limits
Mandelbrot [58] pointed out in his work that the fracture surfaces and objects found in nature, in general, fall into a regular hierarchy, where

different sizes of the irregularities described by fractal geometry, are limited by upper and lower sizes, in which each level is a version in scale of the levels contained below and above of these sizes. Some structures that appear in nature, as opposed to the mathematical fractals, present the property of invariance by scale transformation (self- similarity of self-affinity) only within a limited range of scale transformation ($\varepsilon_{min} \leq \varepsilon \leq \varepsilon_{max}$). Note in Figure 13 that this minimal cutoff scale ε_{min}, one can find an elementary part of the object similar to the whole, that in iteration rules is used as a seed to construct the fractal pattern that is repeated at successive scales, and the maximum cutoff scale D∈R one can see the fractal object as a whole.

One must not confuse this mathematical recursive construction way, with the way in which fractals appear in nature really. In physical media, fractals appears normally in situations of local or global instability [24], giving rise to structures that can be called fractals, at least within a narrow range of scaling ($\varepsilon_{min} \leq \varepsilon \leq \varepsilon_{max}$) as is the case of trees such as pine, cauliflower, dendritic structures in solidification of materials, cracks, mountains, clouds, etc. From these examples it is observed that, in nature, the particular characteristics of the seed pattern depends on the particular system. For these structures, it is easy to see that that fractal scaling occurs from the lowest branch of a pine, for a example, which is repeated following the same appearance, until the end size of the same, and vice versa. In the case of a crack, if a portion of this crack, is enlarged by a scale, ε, one will see that it resembles the entire crack and so on, until reaching to the maximum expansion limit in a minimum scale, ε_{min}, in which one can not enlarge the portion of the crack, without losing the property of invariance by scaling transformation (self-similar or self-affinity). As the fractal growth theory deals with growing structures, due to local or global instability situations [24], such scaling interval is related to the total energy expended to form the structure. The minimum and maximum scales limit is related to the minimum and maximum scale energy expended in forming the structure, since it is proportional to the fractal mass. The number of levels scaling, k, between ε_{min} and ε_{max} depends on the rate at which the formation energy of fractal was dissipated, or also on the instability degree that gave rise to the fractal pattern.

The Fractal Geometric Pattern of a Fracture and Its Measurement Scales

Considering that the fracture surface formed follows a fractal behavior necessarily also admits the existence of a geometric pattern that repeats itself, independent of the scale of observation. The existence of this pattern also shows that a certain degree of geometric information is stored in scale, during the crack growth. Thus, for each type of material can be abstract a kind of geometric pattern, apparently irregular with slight statistical variations, able to describe the fracture surface.

Moreover, for the same type of material is necessary to observe carefully the enlargement or reduction scales of the fracture surface. For, as it reduces or enlarges the scale of view, are found pattern and structures which are modified from certain ranges of these scales. This can be seen in Figure 14. In this figure is shown that in an alumina ceramic, whose ampliation of one of its grains at the microstructure reveals an underlying structure of the cleavage steps, showing that for different magnifications the material shows different morphologies of the surface of fracture.

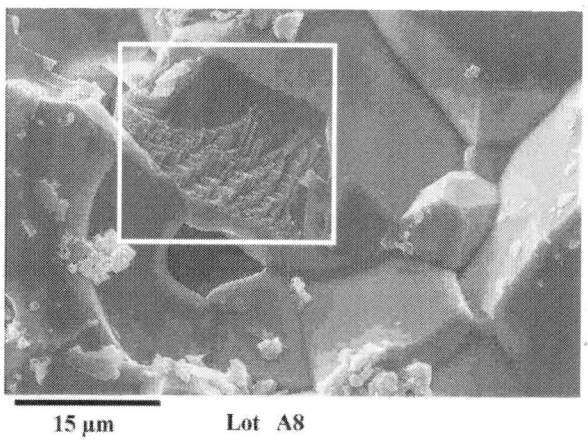

15 μm   Lot A8

Figure 14. Changings in pattern of irregularities with the magnification scale on a ceramic alumina, Lot A8 [56].

To approach this problem one must first observe that, what is the structure for a scale becomes pattern element or structural element to another scale. For example, to study the material, the level of atomic dimensions, the atom that has its own structure (Figure 15a) is the element of another upper level, i.e., the crystalline (Figure 15b). At this level, the cleavage steps formed by the set of crystalline planes displaced, in turn, become the structural elements of microsuperfície fracture in this scale (Figure 15c). At the next level, the crystalline, is the microstructural level of the material, where each fracture microsurface becomes the structural member, although irregular, of the macroscopic rugged fracture surface, as visible to the naked eye, as is shown diagrammatically in Figure 15d.

Thus, the hierarchical structural levels [69] are defined within the material (Figure 15), as already described in this section.

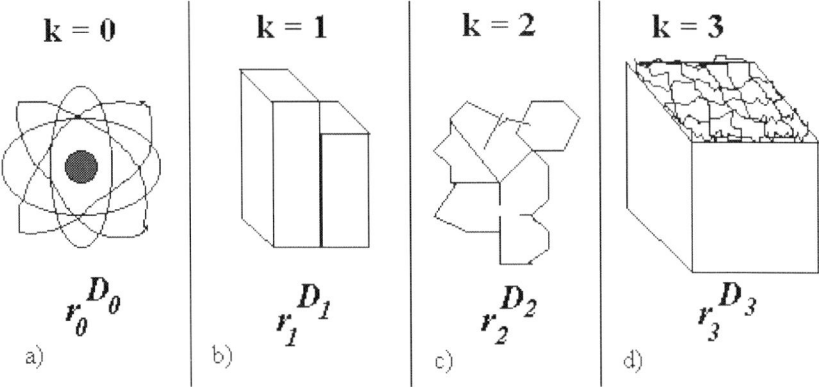

Figure 15. Different hierarchical structural levels of a fracture in function of the observation scale; a) atomic level; b) crystalline level (cleavage steps); c) microstructural level (fracture microsurfaces) and d) macrostructural level (fracture surface).

Based on the observations made in the preceding paragraph, it is observed that the fractal scaling of a fracture surface should be limited to certain ranges of scale in order to maintain the mathematical description of the same geometric pattern (atom, crystal, etc.), which is shown in detail in section - 4.2.3. Although it is possible to find a structural element, forming a pattern, at each hierarchical level, it should be remembered that each type of structure has a characteristic fractal dimension. Therefore, it is impossible to characterize all scale levels of a fracture with only a single fractal dimension. To resolve this problem one can uses a multifractal description. However, within the purpose of this section a description monofractal provides satisfactory results. For this reason, it was considered in the first instance, that a more sophisticated would be unnecessary.

Considering the analytical problem of the fractal description, one must establish a lower and an upper observation scale, in which the mathematical considerations are kept within this range. These scales limits are established from the mechanical properties and from the sample size, as will be seen later. Obviously, a mathematical description at another level of scale, should take into account the new range of scales and measurement rules within this other level, as well as the corresponding fractal dimension.

As already mentioned, the description of the rough fracture surface can be performed at the atomic level, in cleavage steps level (crystalline) or in microstructural level (fracture microsurfaces), depending on the phenomenological degree of detail that wants to reach. This section will be fixed at the microstructural level (micrometer scale), because it reflects the morphology of the surface described by the thermodynamic view of the fracture. This means that the characteristics lengths of generated defects are large in relation to the atomic scale, thus defining a continuous means that reconciles in the same scale the mechanical properties with the thermodynamic properties. Meanwhile, the atomic level and the level of cleavage steps is treated by molecular dynamics and plasticity theory, respectively, which are part areas.

The Calibration Problem of a Fracture Minimum Size as a "Minimal Ruler Size" of Their Fractal

To answer the previous question, about the minimum fracture size, Mishnaevsky [70] proposes a minimum characteristic size, a, given by the size of the smallest possible microcrack, formed at the crack tip (or notch) as a result of stress concentration in the vicinity of a piling up dislocations in the crystalline lattice of the material, satisfying a condition of maximum constriction at the crack tip, where:

$$a \sim k_0 nb, \tag{38}$$

where k_0 is a proportionality coefficient. n is the number of dislocations piling up that can be calculated by:

$$n = \frac{\pi l \sigma (1 - \nu)}{b \mu}, \tag{39}$$

where ν is the Poisson's ratio, l is the length of the piling up of dislocations, σ is the normal or tangential stress, μ is the shear modulus and $\vec{b}$ is the Burgers vector. Substituting (39) in (38) one has;

$$a \sim \frac{k_0 n \pi l \sigma (1 - \nu)}{\mu}. \tag{40}$$

Mishnaevsky equates with mathematical elegance, the crack propagation as the result of a "physical reaction" of interaction of a crack size, $\langle L0 \rangle$ with a piling up of dislocations, nb, forming a microtrinca size, a, i.e.;

$$\langle L_o \rangle + \langle nb \rangle \quad \rightarrow \quad \langle L_o + a \rangle, \tag{41}$$

where $a \ll Lo_e$

$$nb \ll L_o \tag{42}$$

Mishnaevsky proposes a fractal scaling for the fracture process since the minimum scale, given by the size a, until the maximum scale, given by the macroscopic size crack, L_0.

As a consequence for the existence of a minimum fracture size, recently has arosen a hypothesis that the fracture process is discrete or quantized (Passoja, 1988, Taylor et al., 2005; Wnuk, 2007). Taylor et al. (2005) conducted mathematical changes in CFM to validate this hypothesis. Experimental results have confirmed that a minimum fractures length is given by:

$$l_0 \sim \frac{2}{\pi} \left(\frac{K_c}{\sigma_0} \right). \tag{43}$$

where Kc is the fracture toughness, σ_0 is the stress of the yielding strength before the material fracture.

Fractal Scaling of a Self-Similar Rough Fracture Surface or Profile

A mathematical relationship between the extension of the self-similar contour and a extension of its projection is calculated as follows.

Being A the surface extension of the fractal contour, given by a self-similar homogenous function with fractional degree, D, where:

$$A(\varepsilon\delta) = \varepsilon^D A_u(\delta). \tag{44}$$

A0 is the plane projection extension, given by a self-similar homogeneous function with integer degree, d, in accordance with the expression:

$$A_0(\varepsilon\delta)=\varepsilon^d A_u(\delta) \tag{45}$$

where, $A_u(\delta)=\delta d$ is the unit area of measurement, whose values on the rugged and plane surface are the same. Thus the relationships (43) and (44) can be written in the same way as the equations (43) and (44). Therefore, by dividing these equations, one has:

$$A(\varepsilon\delta)=A_0(\delta)\varepsilon^{d-D}. \tag{46}$$

An illustration of the relationship (43), (44) and (45) can be seen in Figure 16.

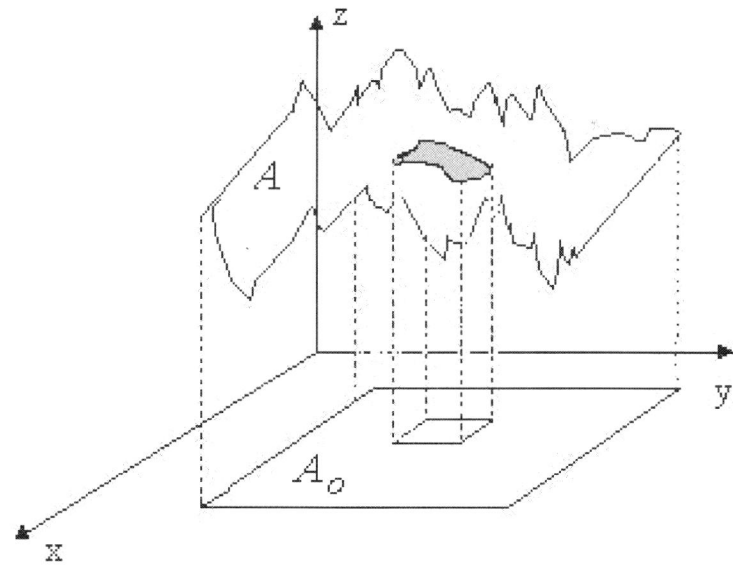

Figure 16. Rugged surface formed by a homogeneous function A, with frationaldegee D, whose planar projection, A0 is a homogeneous function of integer degree d, showing the unit surface area Au.

The rugged fracture surface, may be considered to be a homogeneous function with frational degree, D, ni.e.:

$$A=A_k\varepsilon_k^{-D},\tag{47}$$

and its planar projection, may be considered as a homogeneous function with integer degree d=2, i.e.:

$$A_0=A_r\varepsilon_r^{-d}.\tag{48}$$

The index k was chosen to designate the irregular surface at a k-level of any magnification or reduction. The index r has been chosen to designate the smooth (or flat) surface at a r-level, and the index, 0, was chosen to designate the projected surface corresponding to rugged surface, at the k-level.

Considering that, for k=r and εk=εr, the area unit, Ak and Ar, are necessarily of equal value and dividing relationships (46) and (47), one has:

$$A(\varepsilon k)=A_0\varepsilon_k^{d-D}.\tag{49}$$

The equation (48), means that the scaling performed between a smooth and another irregular surface, must be accompanied by a power term of type ε_k^{d-D}. Thus, there is the fractal scaling, which relates the two fracture surfaces in question: a rugged or irregular surface, which contains the true area of the fracture and regular surface, which contains the projected area of the fracture.

From now on will be obtained a relationship between the rugged and the projected profile of the fracture in analogous way to equation (250) for a thin flat plate (Figure 17a and Figure 17b) with thickness e→0. In this case the area of rugged surface can be written as:

$$A=Le,\tag{50}$$

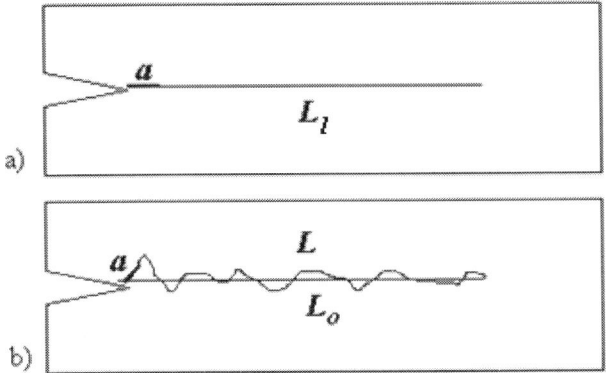

Figure 17. Scaling of a rugged profile of a fracture surface or a crack, using the Mishnaevsky minimum size as a "measuring ruler"; a) in the case of a crack is a non-fractal straight line, where $D=d=1$; b) in the case of tortuous fractal crack, with its projected crack length, where $d \leq D \leq d+1$.

and the area of the projected surface as

$$A0 = L_0 e, \tag{51}$$

According to the equation (48) the valid relationship is:

$$L(\varepsilon k) = L_0 \varepsilon_k^{d-D}, \tag{52}$$

where,

$L(\varepsilon_k)$ is the measured crack length on the scale ε_k, L_0 is the projected crack length measured on the same scale, in a growth direction.

The Self-Similarity Relationship of a Fractal Crack

The fracture is characterized from the final separation of the crystal planes. This separation has a minimum well-defined value, possibly given by theory Mishnaevsky Jr. (1994). If it is considered that below of this minimum value the fracture does not exist, and above it the crack is defined as the crystal planes moving continuously (and the formed crack tip penetrates the material), so that an increasing number of crystal planes are finally separated. One can in principle to use this minimum microscopic size as a kind of ruler (or scale) for the measurement of the crack as a whole([3] -), i.e. from the start point from which the crack grows until its end characterized by instantaneous process of crack growth, for example.

The above idea can be expressed mathematically as follows:

$$L = L_0 \varepsilon^{d-D}, \tag{53}$$

dividing the entire expression (52) above by the minimum Mishnaevsky size one has:

$$\frac{L}{a} = \left(\frac{L_0}{a}\right) \varepsilon^{d-D}, \tag{54}$$

or

$$N = N_0 \varepsilon^{d-D}, \tag{55}$$

Where

$N = L/a$: is the number of crack elements a on the non-projected crack

$N_0 = L_0/a$: is the number of cracl elements a on the crack projectedand yet:

$$\varepsilon = a/L_0, \tag{56}$$

where:

ε: is the scaling factor of the fractal crackd: is the Euclidean dimension of the crack projection D: is the crack fractal dimension.

Within this context the number of microcracks that form the macroscopic crack is given by:

$$N = \left(\frac{a}{L_o}\right)^{-D}. \tag{57}$$

In this context (in Mishnaevsky model), the above expression is volumetric and admits cracks branching generated in the fracture process with opening and coalescence of microcracks. However, he continue equating the process in a one-dimensional way reaching an expression for the crack propagation velocity. A complete discussion of this subject, using a self-affine fractal model to be more realistic and accurate, can be done in another research paper.

The answer to the question about what should be the best scale to be used for fractal fracture scaling is then given as follows: being the limit of the crack length Lk in any scale, given by Lk→L (actual size) as well as lk→lmin, the value of the minimum size ruler, l0 it must be equal to the minimum crack size, a([4] -), given by Mishnaevsky [70], through its energy balance for the fracture of a single monocrystal of the microstructure of a material. The physical reason for this choice is because the Mishnaevsky minimum size is determined by a energy balance, from which the crack comes to exist, because below this size, there is no sense speak of crack length. Therefore, the scale that must be considered is given by:

$$\varepsilon_{min} = a/L_0 \qquad (58)$$

where a is given by relation (40).

Therefore, the statistical self-similarity or self-affinity of a fracture surface, or a crack is limited by a cutoff lower scale ε_{min}, determined by the minimum critical size, l0=a, and a cutoff upper scale ε_{max}, given by the macroscopic crack length, L0.

In two dimensions, the problem of existence of a minimum scale size (possibly given by the Mishnaevky minimum size), leads to abstraction of a microsurface with minimum area, whose shape will be investigated further, in Appendices, in terms of the number of stress concentrators nearest existing within a material.

Model of Self-Affine Fracture Surface or Profiles

In this section one intend to present the development of fractal models of self-similar surfaces. From a rough fracture surface can be extracted numerous profiles also rough on the crack propagation direction. However, in this section is considered only one profile, which is representative of the entire fracture surface (Figure 18). The plane strain condition admits this assumption. Because, although the fracture toughness varies along the thickness of the material to a plastic zone reduced in relation to material thickness, it can be considered a property. This means that it is possible to obtain a statistically rough profile, equivalent to other possible profiles, which can be obtained within the thickness range considered by plane strain conditions.

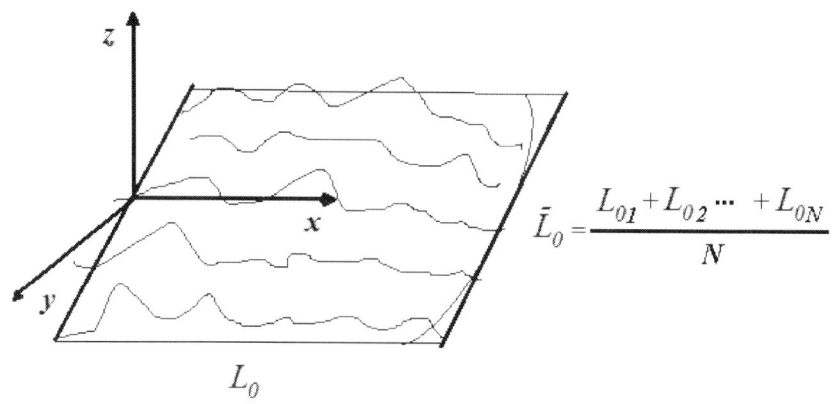

Figure 18. Statistically equivalent profiles along the thickness of the material

In order also equivalent to this, it is also possible to obtain an average projected crack length as a result of an average of the crack size along the thickness of the material thickness within the range considered by plane strain, for the purpose of calculations in CFM, it is considered this average size as if it were a single projected crack length, as recommended by the ASTM – E1737-96 [71]. Therefore, inwhat follows, is effected by reducing or lowering the dimensional degree of relationship from the two-dimensional case, shown above, for the one-dimensional case, as follows:

$$A(x,y) \rightarrow L(x). \tag{59}$$

thus, for a self-affine fractal one has:

$$L(\lambda_x x) = \lambda_x{}^H L(x), \tag{60}$$

where

$$H = 2 - D \qquad\qquad (61)$$

is the Hurst exponent measuring the profile ruggedness. In one-dimensional case the fracture surface is a profile whose length L is obtained from measuring the projected length, L_0, as illustrated below in Figure 19.

Calculation of the Rugged Crack Length as a Function of its Projected Length

Considering a profile of the fracture surface as a self-affine fractal, analogous to the fractal of Figure 19, which perpendicular directions have the same physical nature the Voss [48] equation to the Brownian motion can be generalized([5] -) to obtain rugged crack length L, depending on the projected crack length, L0.

Figure 19 illustrates one of the methods for fractal measuring. This measure can be obtained by taking boxes or rectangular portions, based ΔL0 and height ΔH0 on the crack profile, and recovering up this profile, within these boxes, with "little boxes" (recovering units) with small sizes, l0 and h0, respectively (Figure 19). Instead of little boxes is also possible to use other shapes([6] -) compatible with the object to be measured. Then makes the counting of the little boxes (or recovering units) needed to recover the extension of the rugged crack, centered in the box ΔL0×ΔH0. The number of these little boxes (or recovering units) of size r in function of the boxes extension (or parts), ΔL0×ΔH0, provides the fractal dimension, as shown in section 3 - Methods for Measuring Length, Area, Volume and Fractal Dimension.

Assume that the rectangular little boxes (or recovering units) of microscopic size, r, recover the entire crack length, ΔL inside the box with greater length, ΔL0×ΔH0. The number of little boxes (recovering unit) with sides of l0×h0 needed to recover a crack in the horizontal direction, inside the box (or stretch) of rectangular area ΔL0×ΔH0, for the self-affine fractal can be obtained by the expression:

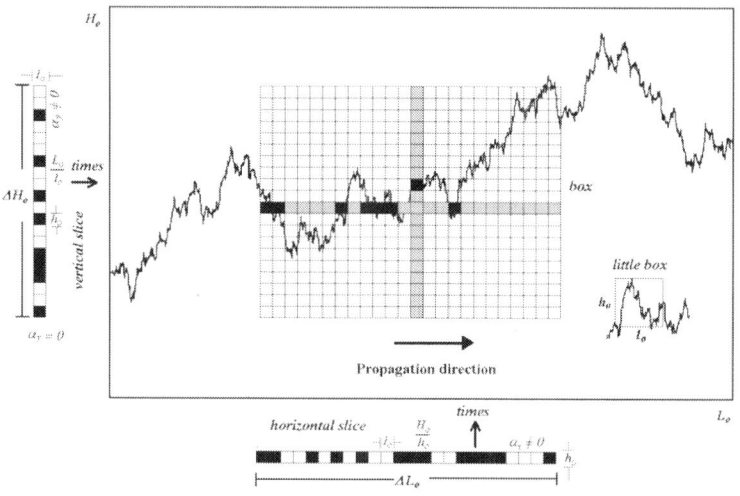

Figure 19. Self-affine fractal of Weierstrass-Mandelbrot, where $\varepsilon k=1/4$ and Dx=1.5 and H=0.5, used to represent a fracture profile (Family, Fereydoon; Vicsek, Tamas Dynamics of Fractal Surfaces, World Scientific, Singapore, 1991, p.7).

$$N_v = \frac{\Delta L_0}{l_0} \varepsilon_v{}^0 \text{ (in vertical direction)}$$

(62)

where ΔL_0 is the crack horizontal projection and εv is the vertical scaling factor.

Considering that the self-affine fractal extends in the horizontal direction along L0, and oscillates in the perpendicular direction, i.e. in the vertical direction, the number of little boxes Nh, with size, l0 in the horizontal direction, are gathered to form the projected length L0, while vertically the number of little boxes Nv, with size h0, overlap each other, increasing (as power law) this number in comparison to the number of little boxes gathered horizontally. Therefore, for the vertical direction with a projection $\Delta H0$, the box sides $\Delta L_0 \times \Delta H_0$, an expression for the number of boxes (or units covering) can be writen as:

$$N_h = \frac{\Delta H_0}{h_0} \varepsilon_h{}^{-H} \text{ (in horizontal direction)} .$$

(63)

where H is the Hurst exponent, $\Delta H0$ is the total variation in height ($lo \leq \Delta H0 \leq \Delta Lo$) and εh is the scale transformation factor in the horizontal direction.

Therefore, for the corresponding rugged crack length (real) ΔL, the stretch $\Delta L0 \times \Delta H0$ one can writes:

$$\Delta L = N_v r \tag{64}$$

where r is equal to the rugged crack length on a microscopic scale, as a function of extension of the little boxes $l0 \times h0$ by:

$$r = \sqrt{l_0{}^2 + h_0{}^2} \tag{65}$$

where l0 and h0 are the microscopic sizes of the crack length in horizontal and vertical directions, respectively. Substituting (64) in (63), one has:

$$\Delta L = N_v \sqrt{l_0^2 + h_0^2} \tag{66}$$

substituting (61) in (65), one has:

$$\Delta L = \frac{\Delta L_0}{l_0} \sqrt{l_0{}^2 + h_0{}^2} \tag{67}$$

Since that in the fracture process, the scales in orthogonal directions are the same physical nature, one can choose $\varepsilon v = \varepsilon h = l0/\Delta L0$, and one can writes from (62) that:

$$N_h = \left(\frac{\Delta H_0}{l_0}\right)\left(\frac{\Delta L_0}{l_0}\right)^H \tag{68}$$

being necessarily

Nh=Nv, one has:

$$\left(\frac{\Delta L_0}{l_0}\right) = \left(\frac{\Delta H_0}{l_0}\right)\left(\frac{\Delta L_0}{l_0}\right)^H \tag{69}$$

rewriting the equation (66), one has:

$$\Delta L = \Delta L_0 \sqrt{1 + \left(\frac{h_0}{l_0}\right)^2}$$

(70)

writing h_0 from (68), as:

$$h_0 = \Delta H_0 \left(\frac{\Delta L_0}{l_0}\right)^{H-1}.$$

(71)

Eliminating in (69) the dependence of h0, by substituting (70) in (69), one has:

$$\Delta L = \Delta L_0 \sqrt{1 + \left(\frac{\Delta H_0}{l_0}\right)^2 \left(\frac{\Delta L_0}{l_0}\right)^{2(H-1)}}$$

(72)

The curve length in the stretch, $\Delta L0 \times \Delta H0$ considering the Sand-Box method [38] whose counting starts from the origin of the fractal, can be written as:

$\Delta L{=}L, \Delta L0{=}L0$ and $\Delta H0{=}H0$ hence the equation (71) shall be given by:

$$L = L_0 \sqrt{1 + \left(\frac{H_0}{l_0}\right)^2 \left(\frac{L_0}{l_0}\right)^{2(H-1)}},$$

(73)

whose the plot is shown in Figure 20. Note that the lengths

L0 and H0 correspond to the projected crack length in the horizontal and vertical directions, respectively.

Applying the logaritm on the both sides of equation (72) one obtains an expression that relates the fractal dimension with the projected crack length:

$$D_f \equiv \frac{\ln(L/l_0)}{\ln(L_0/l_0)} = 1 + \frac{1}{2} \frac{\ln\left\{\left(\frac{H_0}{l_0}\right)^2 \left(\frac{L_0}{l_0}\right)^{2(H-1)}\right\}}{\ln L_0}$$

(74)

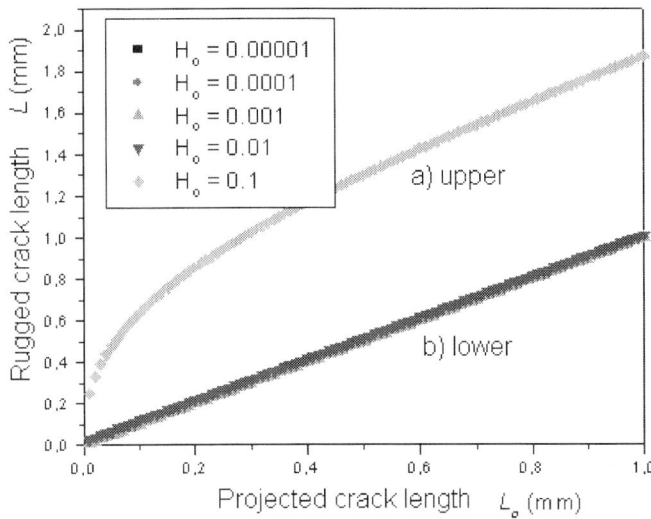

Figure 20. Graph of the rugged length L in function of the projected length L0, showing the influence of height, H0, of the boxes in the fractal model of fracture surface: a) in the upper curves is observed the effect of H0 as it tends to unity (H0→1.0), b) in the lower curves, that appearing almost overlap, is observed the effect of H0 as it tends to zero (H0→0).

The graph in Figure 20 shows the influence of the boxes height H0 on the rugged crack length, L, as a function of the projected crack length, L0. Note that for boxes of low height (H0→0), in relation to its projected length, L0, the lower curves (for H0=0.01,0.001,0.0001), denoted by the letter "b", almost overlap giving rise to a linear relation between these lengths (Figure 21). While for boxes of high height (H0→1.0) in relation to its projected length, L0, the relation between the lengths become each more distinct from the linear relationship for the same exponent roughness, H.

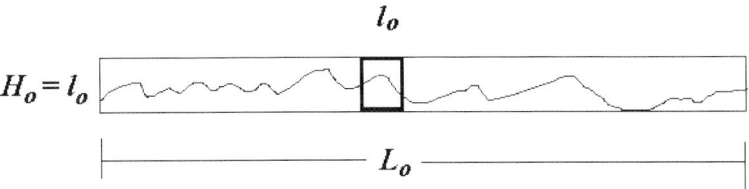

Figure 21. Counting boxes (or strechts) with rectangular sizes

LoxHo where the boxes that recovers the profile have different extensions in the horizontal and vertical directions.

Making up the counting boxes (or stretch) with rectangular sizes Lo x Ho where the boxes recovering the profile have different extensions in the horizontal and vertical directions respectively, i.e., Ho = lo the equation (72). Is simplified to:

$$L = L_o \sqrt{1 + \left(\frac{l_o}{L_o}\right)^{2H-2}}.$$

(75)

which plot is shown in Figure 22.

The graph in Figure 22 shows the influence of the roughness dimension on the rugged crack length, L, in function of the projected length, L0. Note that for H0→1.0, corresponding to a smooth surface, the relation between the rugged and projected length becomes increasingly linear. While for H0→0, which corresponds to a rougher surface, the the relation between the rugged and projected length becomes increasingly non-linear.

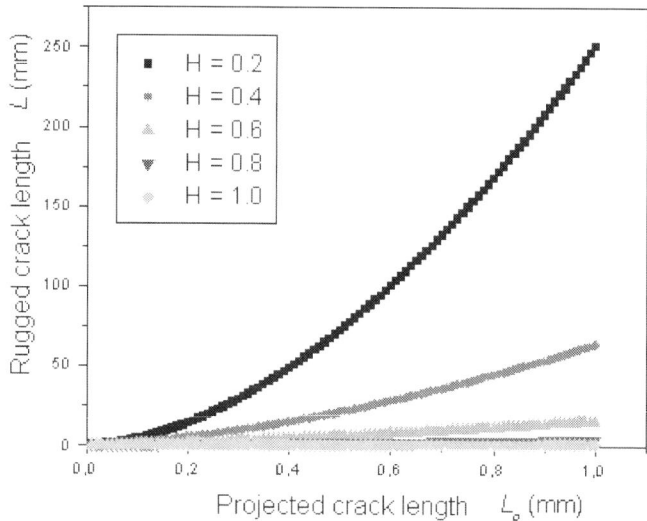

Figure 22. Graph of the rugged length, L, in function of the projected length, L0, showing the influence of the Hurst exponent H, in the fractal model of the fracture surface.

Note that for Lo=Ho, one has, from the equation (62) and (68) the following relationship:

$$L_O = h_o \left(\frac{l_o}{L_o} \right)^{-H} ,$$

(76)

which is a self-similar relation between the projected crack length, L0, and height of the little box, h0. This relationship shows that all self-affine fractal, in the approximation of a small scale, has a local self-similarity forming a fractal substructure, when is considered square portions, L0×L0, instead of rectangular portions, L0×H0.

It important to observe that L0 denotes the distance between two points of the crack (the projected crack length). The self-affine measure, L of L0, in the fractal dimension, D, is given by (72). l0 is the possible minimum length of a micro-crack, which defines the scale l0/L0 under which the crack profile is scrutinized, as discussed in previous section and will be discussed after in the section #.5.4.5. The Hurst exponent, H, is related to D by (60).

In the study of a self-affine fractal there are two extremes limits to be verified. One is the limit at which the boxes height is high in relation to its projected length, L0, i.e. (H0→L0), which is also called local limit. The other limit is one in which the boxes height is low in relation to its projected length, L0, i.e., (H0→h0) which is called global limit. It will be seen now each one of this limits case contained in the expression (72).

Case. 1 : The self-similar or local limit of the fractality

Taking the local limit of the self-affine fractal measure as given by (72), i. e. for the case where,

H0=L0>>l0, one has:

$$L \cong L_o \left(\frac{l_o}{L_o} \right)^{H-1}$$

(77)

where

$$\frac{L}{L_o{}^{2-H}} \cong l_o{}^{H-1} = constant$$

(78)

This equation is analogous to self-similar mathematical relationship only that the exponent is $(1-H)$ instead of $(D-1)$, which satisfies the relation $H=2-D$ [3, 40, 51, 70].

According to these results it is observed that the relation (77) has a commitment to the Hurst exponent of the profiles on the considered observation scale $\varepsilon=l0/L0$. It is observed that the consideration of a minimum fracture size $l01$ over a region, one must consider the local dimension of the fracture roughness on this scale. Similarly, if the considerations of a minimum fracture size are made in a scale that involves several regions, $l02$ this should take into account the value of the roughness global dimension on this scale, so that:

$$(2 - H_1)l_{01}^{H_1-1} = (2 - H_2)l_{02}^{H_2-1} = constant,$$

(79)

although $l01 \neq l02$ e $H1 \neq H2$

Case. 2: The self-affine or global limit of fractality

Taking the global limit of the self-affine fractal measure given by (72), i.e. for the case in which: $H0=l0<<L0$. Therefore the length L is independently of H and $D=1$, so

$$L \cong Lo$$

(80)

It must be noted that the ductile materials by having a high fractality have a crack profile which can be better fitted by the equation (76), while brittle materials by having a low fractality will be better fitted by the equation (79) corresponding the classical model, i.e., a flat geometry for the fracture surface. Furthermore, the cleavage which occurs on the microstructure of ductile materials tend to produce a surface, where $L \cong L0$, which could be called smooth. However, this cleavage effect is just only local in these materials and therefore the resulting fracture surface is actually rugged.

Local Ruggedness of a Fracture Surface
Defining the local roughness of a fracture surface, as:

$$\xi = \frac{dA}{dA_o} \Rightarrow A = \int \xi(A_0)\, dA_0.$$

(81)

where A is the rugged surface and A_0 is the projected surface. In the case of a rugged crack profile, one has:

$$\xi \equiv \frac{dL}{dL_o} \Rightarrow L = \int \xi\left(L_0\right) dL_0$$

(82)

using (74) in (81), one has that:

$$\xi \equiv \frac{1 + (2 - H)\left(\frac{l_o}{L_o}\right)^{2H-2}}{\sqrt{1 + \left(\frac{l_o}{L_o}\right)^{2H-2}}}$$

(83)

From (81) note that when there is no roughness on surfaces (flat fracture) one has that: L=L0, thus

$$\frac{dL}{dL_o} = 1.$$

(84)

The quantity $\xi \equiv dL/dLo$ seems be a good definition of ruggedness unlike the definition where the ruggedness is given by $\xi = L/L//$ [56, 57] (where L//=L0Mcosθ, see Figure 23) does not satisfy the requirement intuitive of the ruggednes when L0M is only inclined with respect to L0, while maintaining,

L0=L0M, as shown Figure 23.

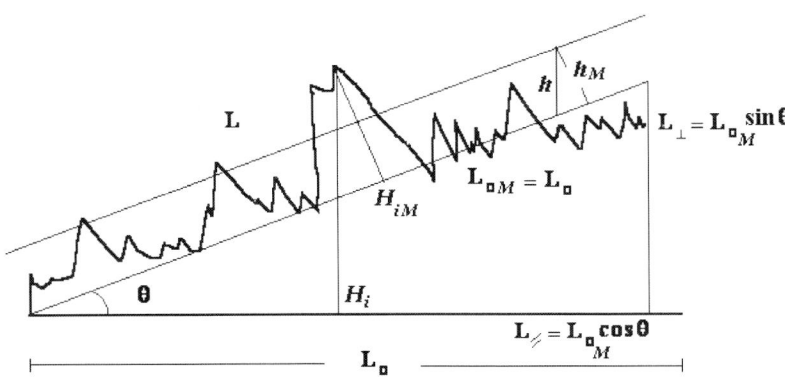

Figure 23. Schematization of a rugged surface which is inclined with respect to its projection.

The ruggedness must depend on infinitesimally of the projected length and its relative orientation to it. In this case, the surface roughness by the usual definition adds an error equal to the angle secant θ, or

$$\xi = L/L_{//} = \left(\frac{L}{L_{0M}}\right)\left(\frac{L_{0M}}{L_{//}}\right) = \left(\frac{L}{L_{0M}}\right)\frac{1}{\cos\theta}.$$

(85)

However, by the definition proposed herein, when only one inclines a smooth surface against to the horizontal, one has: L=L0M and L0M=L0 and again, $\xi = \frac{dL}{dL_o} = \frac{dL}{dL_{oM}}\frac{dL_{oM}}{dL_o} = 1.$

Within this philosophy will be considered as rugged any surface that presents in an infinitesimal portion a variation of their contour such that

dL/dLo>1, therefore has to be:

$$dL/dLo \geq 1$$

(86)

Preliminary Considerations on the Proposed Model

Considering a fractal model for the fracture surface given by the equation:

$$\Delta L = \Delta L_o \sqrt{1 + \left(\frac{\Delta H_o}{l_o}\right)^2 \left(\frac{l_o}{\Delta L_o}\right)^{2H-2}},$$

(87)

it is possible to describe its ruggedness in order to include it in the mathematical formalism CFM to obtain a Fractal Fracture Mechanics (FFM). This derivative of equation (86) defines a fractal surface ruggedness, which for the case of a self-affine crack which grows with,

$\Delta H0 \rightarrow l0$, is given by:

$$\xi = \frac{1 + (2 - H)\left(\frac{l_o}{\Delta L_o}\right)^{2H-2}}{\sqrt{1 + \left(\frac{l_o}{\Delta L_o}\right)^{2H-2}}} \geq 1.$$

(88)

such modifications were added to equations of the Irregular Fracture Mechanics to obtain a Fractal Fracture Mechanics as described below.

Comparison of Fractal Model With Experimental Results

In Figure 24 and Figure 25, a good agreement is observed in the curve fitting of equation (72) and equation (73) to the fractal analyses of the mortar specimen A2 side 1 and the red ceramic specimen A8, respectively.

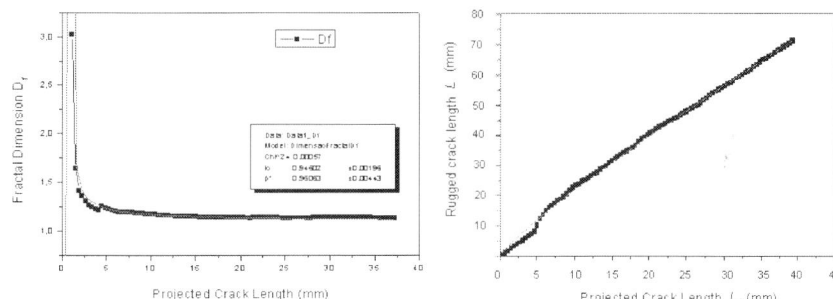

Figure 24. a) Fractal analysis of mortar specimen A2 side 1 – Fractal dimension x Projected length, L0; **b)** Fractal analysis of mortar specimen A2 side 1 - rugged length L x projected length, L0

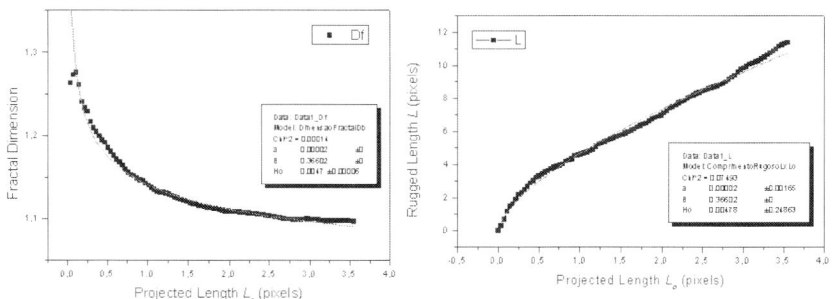

Figure 25. a) Fractal analysis of red clay A8 side 1 – Fractal dimension x Projected length, L0; **b)** Fractal analysis of red ceramic specimen A8 - rugged length L x projected length, L0

CONCLUSIONS

1. It is possible, in principle, mathematically distinguish a crack in different materials using geometric characteristics which can be portrayed by different values of roughness exponents in the relations (72) and (74).
2. The fractal model of the rugged crack length, L in function of the projected crack length, L0, suggested by Alves [72 73,74, 75] seems have a good agreement with experimental results. This results allowed us to consolidate the model previously published in the literature on fracture [72,73, 75].
3. The rugged crack length is a response to its interaction with the microstructure. of the material. Therefore, mathematically is possible to portray the rugged peculiar behavior of a crack using fractal geometry
4. The mathematical model presents a wealth (mathematical richness) that can still be explored in terms of determining the minimum crack length, l0 for each material and the fractal dimension as a function of test parameters and material properties.
5. The mathematical model is sensitive to variations in the behavior of the crack length it is a linear or logarithmic with the projected crack length.

Comparing the experimental results with the model proposed in this chapter, it is concluded that one of the more important results obtained here are the equations (72), (74) and (82) leading to finding that the fracture surfaces of the materials analyzed are indeed (actually) self-affine fractals. Starting from this verification it becomes feasible to consider the fractal model of rugged fracture surface and its ruggedess inside the equations of the classical fracture mechanics, according to equation (74) and (82). As there is a close relationship between phenomenology and structure formed by virtue of its fractal geometry, the understanding of the formation processes of these dissipative structures, as the cracks, should be derived from their mathematical analysis, as the close relationship between the phenomenology of the formation process of dissipative structures and their fractal geometry. Therefore, the mathematical description of fractal structures must exceed a simple geometrical characterization, in order to correlate the pattern formed in the process of energy dissipation with the amount of energy dissipated in the process that generated it. Thus, it is possible to use the fractal geometry in order to understand other more and more complex processes inside the fracture mechanics. Therefore, the various mechanisms responsible by the crack deviation and by the formation of the rugged fracture surface can then, from the fractal model, be quantified in the fractal analysis of this surface.

The idea of obtaining a relationship between L and L0 comes the need to maintain the present formalism used by the CFM, showing that fractal geometry can greatly contribute to the continued advancement of this science.

On the other hand, we are interested in developing a Fractal Thermodynamic for a rugged crack that will be related to the CFM and the Classical Fracture Thermodynamics when the crack ruggedness is neglected or the crack is considered smooth.

NOTES

- Voss present a fractal description for the noise in the Brownianomouvement
- This must be done so that the measurement scales are not arbitrary and may depend on some property of the material.
- During or concurrently with its propagation, in a dynamic scaling process, or not
- It is possible that this minimal ruler size be very low than the scale used in fractal characterization of the fracture surfaces. However, it must to be the smallest possible size for a microcrack.
- Voss [48], modeled the noise plot of the frational Brownian motion , where in the y-direction, he plots the amplitude, VH, and in the x-direction, he plots the time, t.
- Some authors used "balls"

REFERENCES

1. MandelbrotBenoit B. (1982The Fractal Geometry of NatureSan Francisco, Cal-Usa, New York: W. H. Freeman and Company. 472 p.
2. J. A Rodrigues, E V. C Pandolfelli, 1996Dimensão Fractal E Energia Total de Fratura. Cerâmica, Maio/Junho. 42(275).
3. F. M Borodich, 1997Some Fractals Models of Fracture. J. Mech. Phys. Solids. 452239259
4. HepingXie (1989The Fractal Effect of Irregularity of Crack Branching on the Fracture Toughness of Brittle MaterialsInternational Journal of Fracture41267274
5. Z. Q Mu, and C. W Lung, 1988Studies on the Fractal Dimension and Fracture Toughness of SteelJ. Phys. D: Appl. Phys. 21848850

6. J. J Mecholsky, D. E Passoja, and K. S Feinberg-ringel, 1989Quantitative Analysis of Brittle Fracture Surfaces Using Fractal GeometryJ. Am. Ceram. Soc. 7216065

7. G. M Lin, J. K. L Lai, 1993Fractal Characterization of Fracture Surfaces In a Resin-Based CompositeJournal of Materials Science Letters12470472

8. NagahamaHiroyuki (1994A Fractal Criterion for Ductile and Brittle FractureJ. Appl. Phys. 15 March, 75632203222

9. LeiWeisheng and Bingsen Chen (1994Discussion on "The Fractal Effect of Irregularity of Crack Branching on the Fracture Toughness of Brittle Materials" By XieHepingInternational Journal of FractureR65R70.

10. LeiWeisheng and Bingsen Chen (1994Discussion: "Correlation Between Crack Tortuosity and Fracture Toughness In CementitiousMaterial"By M. A. Issa, A. M. Hammad and A. Chudnovsky, A. International Journal of FractureR29R35.

11. LeiWeisheng and Bingsen Chen (1995Fractal Characterization of Some Fracture PhenomenaEngineering Fracture Mechanics502149155

12. M Tanaka, 1996Fracture Toughness and Crack Morphology In Indentation Fracture of Brittle MaterialsJournal of Materials Science31749755

13. T Chelidze, Y Gueguen, 1990Evidence of Fractal FractureTechnical Note) Int. J. Rock. Mech Min. Sci&GeomechAbstr. 273223225

14. H. J Herrmann, J Kertész, L De Arcangelis, 1989Fractal Shapes of Deterministic CracksEurophys. Lett. 102147152

15. L De Arcangelis, Hansen A; Herrmann, H. J. (1989Scaling Laws In Fracture. Phys. Review B. 1 July 40(1).

16. HerrmannHans J. (1986Growth: An Introduction. In: H. Eugene Stanley and Nicole Ostrowskym editors. On the Growth and Form, Fractal and Non-Fractal Patterns In Physics, NatoAsi Series, Series E: Applied Sciences N. 100 (1986), Proc. of the Nato Advanced Study Institute Ön Growth and Form", Cargese, Corsiva, France June 27July 6 1985. Copyright ByMartinusNighoff Publishers, Dordrecht.

17. C Tsallis, A. R Plastino, and W Zheng, M Chaos, Solitons& Fractals 8: 885.

18. MccauleyJoseph L. (1993Chaos, Dynamics and Fractals: An Algorithmic Approach To Deterministic Chaos,Cambridge Nonlinear Science Series, 2Cambridge England: Cambridge University Press.

19. H Stanley, Eugene (1973Introduction To Phase Transitions and Critical PhenomenaEditors: Cooperative Phenomena Near Phase Transitions, a Bilbiography With Selected Readings, Mit,). Cambridge, Massachusetts : Claredon Oxford

20. D. I Uzunov, 1993Theory of Critical Phenomena, Mean Field, Flutuactions and Renormalization. Singaore: World Scientific Publishing Co. Pte. Ltd.

21. C Beck, and F Schlögl, 1993Thermodynamics of Chaotic Systems: An Introduction,Cambridge Nonlinear Science Series, 4Cambridge England: Cambridge University Press,.

22. MeakinPaul (1995Fractal Growth:, Cambridge Nonlinear Science Series, 5England: Cambridge University Press.

23. VicsékTamás (1992Fractal Growth Phenonmena. Singapore: World Scientific.

24. L. M Sander, 1984Theory of Fractal Growth Process In: F. Family, D. P. Landau editors. Kinetics of Aggregation and Gelation. Amsterdam: Elsevier Science Publishers B. V. 1317

25. MeakinPaul (1993The Growth of Rough Surfaces and InterfacesPhysics Reports. December 235485189289

26. L Pietronero, AErzan, C Everstsz, 1988Theory of Fractal GrowthPhys. Revol. Lett. 15 August 617861864

27. HerrmannHans J.; Roux, Stéphane (1990Statistical Models For the Fracture of Disordered MediaRandom Materials and Processes Series. H. Eugene Stanley and Etienne Guyoneditors., Amsterdam: North-Holland.

28. J. C Charmet, S Roux, and E Guyon, 1990Disorder and Fracture. New York: Plenum Press.

29. MeakinPaul; Li, G.; Sander, L. M.; Louis, E.; Guinea, F. (1989A Simple Two-Dimensional Model For Crack PropagationJ. Phys. A: Math. Gen. 2213931403

30. GrossSteven. P.; Jay. Fineberg, M. P. Marder, W. D. Mccormick and Harry. L. Swinney (1993Acoustic Emissions From Rapidly Moving Cracks. Physical Review Letters., 8 November 711931623165

31. SharonEran; Steven Paul Gross and Jay Fineberg (1996Energy Dissipation In Dynamic FracturePhysical Review LettersMarch, 761221172120

32. F Hausdorff, 1919Dimension Und ÄußeresMaß. MathematischeAnnalen March 79(1-2): 157-179. Doi:10.1007/Bf01457179.

33. A. S Besicovitch, 1929On Linear Sets of Points of Fractional Dimensions. MathematischeAnnalen

34. A. S. H. D Besicovitch, Ursell. (1937) Sets of Fractional Dimensions. Journal of the London Mathematical SocietySeveral Selections From This Volume Are Reprinted In Edgar, Gerald A. (1993Classics on Fractals. Boston: Addison-Wesley. 0-20158-701-7Chapters 9,10,11

35. MandelbrotBenoit B (1975Fractal..

36. MandelbrotBenoit B (1977Fractals: Form Chance and Dimension. San Francisco, Cal-USA: W. H. Freeman and Company.

37. AlvesLucasMáximo (2012Application of a Generalized Fractal Model For Rugged Fracture Surface To Profiles of Brittle Materials. Paper in preparation.

38. BundeArminShlomoHavlin (1994Fractals InScienceSpringer-Verlag.

39. FamilyFereydoon; Vicsek, Tamás (1991Dynamics of Fractal SurfacesSingapore: World Scientific Publishing Co. Pte. Ltd. 78P. 73-77.
40. FederJens (1989Fractals, New York: Plenum Press.
41. BarnsleyMichael (1988Fractals EverywhereAcademic Press, Inc, Harcourt Brace Jovanovich Publishers.
42. V Milman, Yu.,Blumenfeld R., Stelmashenko N. A. and Ball R. C. (1993Phys. Rev. Lett. 71, 204.
43. MilmanVictor Y.; Nadia A. Stelmashenko and Raphael Blumenfeld (1994Fracture Surfaces: a Critical Review of Fractal Studies and a Novel Morphological Analysis of Scanning Tunneling Microscopy MeasurementsProgreess In Materials Science. 38425474
44. YamagutiMarcos (1992Doctoral Thesis- Universidade de São Paulo
45. AllenMartin; Gareth J. Brown; Nick J. Miles (1995Measurements of Boundary Fractal Dimensions: Review of Current Techniques. Powder Technology84114
46. L. F Richardson, The Problem of Contiguity: An Appendix To Statistics of Deadly Quarrels. General Systems Yearbook. (6): 139-187.
47. AlvesLucasMáximo (1998Escalonamento Dinâmico da FractaisLaplacianosBaseado No Método Sand-Box, In: Anais Do 42o Cong. Bras. de Cerâmica, Poços de Caldas de 3 a 6 de Junho,.ArtigoPublicadonesteCongresso Ref.007/1.
48. VossRichard F. (1991In: Family, Fereydoon. and Vicsék, Tamás, editors. Dynamics of Fractal Surfaces. Singapore: World Scientific. 4045
49. MorelSthéphane, Jean Schmittbuhl, Elisabeth Bouchaud and Gérard Valentin (2000Scaling of Crack Surfaces and Implications on Fracture Mechanics Arxiv:Cond-Mat/0007100 6 Jul 2000, 1Or Phys. Rev. Lett. 21 August 85(8).
50. UnderwoodErwin E. and KingshukBanerji (1996Quantitative Fractography. Engineering Aspectes of Failure and Failure Analysis- Asm- Handbook-12Fractography- the Materials Information Society (1992). Astm 1996, 192209
51. R. H Dauskardt, F Haubensak, and R. O Ritchie, 1990On the Interpretation of the Fractal Character of Fracture Surfaces; Acta Metall. Matter. 382143159
52. Underwood Erwin E. and Kingshuk Banerji (1986 Fractals In Fractography Materials Science and EngineeringEd. Elsevier. 80: 114
53. Underwood Erwin E. and Kingshuk Banerji (1996Fractal Analysis of Fracture Surfaces. Engineering Aspectes of Failure and Failure Analysis-Asm- Handbook- 12Fractography- the Materials Information Society (1992), Astm 1996. 210215
54. L. M Alves, Chinelatto, Adilson Luiz; Chinelatto, Adriana Scoton Antonio; Prestes, Eduardo (2004Verificação de um modelo fractal de fratura de argamassa de cimento. In: Anais do 48° CongressoBrasileiro de Cerâmica, 28 de Junho a 1° de Julho de 2004, Em Curitiba- Paraná.

55. L. M Alves, Chinelatto, Adilson Luiz; Chinelatto, Adriana Scoton Antonio ; Grzebielucka, Edson Cezar (2004Estudo do perfil fractal de fratura de cerâmicavermelha. In: Anais do 48° CongressoBrasileiro de Cerâmica, 28 de Junho a 1° de Julho de 2004, Em Curitiba- Paraná.

56. Dos SantosSergio Francisco (1999Aplicação Do Conceito de Fractais Para Análise Do Processo de Fratura de MateriaisCerâmicos, Master Dissertation, Universidade Federal de São Carlos. Centro de CiênciasExatas e de Tecnologia, Programa de Pós-GraduaçãoemCiência e Engenharia de Materiais, São Carlos.

57. T. L Anderson, 1995Fracture Mechanics, Fundamentals and Applications.Crc Press, 2th Edition.

58. MandelbrotBenoit B.; Dann E. Passoja& Alvin J. Paullay (1984Fractal Character of Fracture Surfaces of MetalsNatureLondon). 19 April, 308 [5961]: 721722

59. C. W Lung, and Z. Q Mu, 1988n Fractal Dimension Measured With Perimeter Area Relation and Toughness of Materials, Physical Review B. 1 December, 38161178111784

60. BouchaudElisabeth (1977Scaling Properties of Crack.J. Phys: Condens. Matter. 943194344

61. E Bouchaud, G Lapasset, and J Planés, 1990Fractal Dimension of Fractured Surfaces: a Universal Value?Europhysics Letters. 1317379

62. E Bouchaud, J. P Bouchaud, 1994I) Fracture Surfaces: Apparent Roughness, Relevant Length Scales, and Fracture Toughness. Physical Review B. 15 December. 50231775217755

63. BouchaudElisabeth (1997Scaling Properties of CracksJ. Phy. Condens. Matter. 943194344

64. A. B Mosolov, 1993Mechanics of Fractal Cracks In Brittle SolidsEurophysics Letters. 10 December, 248673678

65. Family &Vicsek (1985Scaling in steady-state cluster-cluster aggregate. J. Phys. A 18: L75.

66. BarabásiAlbert- László; H. Eugene Stanley (1995Fractal Concepts In Surface GrowthCambridge: Cambridge University Press.

67. LopezJuan M. Miguel A. Rodriguez, and Rodolfo Cuerno (1997Superroughening versus intrinsic anomalous scaling of surfacesPhys. Rev. E 56439933998

68. LopezJuan M. and Schmittbuhl, Jean (1998Anomalous scaling of fracture surfacesPhys. Rev. E. 57664056408

69. A. G Guy, 1986Ciências Dos Materias, Editora Guanabara 435p.

70. MishnaevskyJrL. L. (1994A New Approach To the Determination of the Crack Velocity Versus Crack Length RelationFatigueFract. Engng. Mater. Struct. 171012051212

71. ASTM- E17371996Standard Test Method For J-Integral Characterization of Fracture Toughness. Designation Astm E1737/96,124

72. AlvesLucasMáximo; RosanaVilarim da Silva and Bernhard Joachim Mokross (2000In: New Trends In Fractal Aspects of Complex Systems- FACS 2000IUPAP International Conference October, 16, 2000m At Universidade Federal de Alagoas- Maceió, Brasil.

73. AlvesLucasMáximo; RosanaVilarim da Silva, Bernhard Joachim Mokross (2001The Influence of the Crack Fractal Geometry on the Elastic Plastic Fracture Mechanics. PhysicaAStatistical Mechanics and Its Applications. 12 June 2001, 295, (1/2): 144 EOF148 EOF

74. AlvesLucasMáximo (2002Modelamento Fractal da Fratura E Do Crescimento de TrincasEmMateriais. Relatório de Tese de DoutoradoEmCiência E Engenharia de Materiais, Apresentada À InterunidadesEmCiência E Engenharia de Materiais, da Universidade de São Paulo-Campus, São Carlos, Orientador: Bernhard Joachim Mokross, Co-Orientador: José de Anchieta Rodrigues, São Carlos- SP.

75. AlvesLucasMáximo (2005Fractal Geometry Concerned with Stable and Dynamic Fracture MechanicsJournal of Theorethical and Applied Fracture Mechanics. 44/1:4457

LIST OF ABBRIVATION

DLA	Diffusion Limited Aggregation
BD	Ballistic Deposition
SF	Fracture Surfaces
CFM	Classical Fracture Mechanics

CHAPTER 2

Ways of Reducing the Impact of Mechanical Vibrations on Hydraulic Valves

M. Stosiak[1,2]

[1]Wrocław University of Technology, Faculty of Mechanical Engineering, 50-371 Wrocław, Poland
[2]Division of Hydraulic Machines and Systems, Institute of Machines Design and Operation, Wrocław University of Technology, Poland

ABSTRACT

This paper deals with the impact of mechanical vibrations on the environment, particularly on hydraulic valves. The main sources of such vibrations and their effects on hydraulic systems are indicated. Some documents setting down standard requirements for resistance to vibrations and to the noise generated by vibrations are cited. Two ways of reducing the impact of mechanical vibrations on the valve are proposed and a theoretical analysis, constituting the basis for selecting a material for an effective vibration isolator for the valve, is carried out.

INTRODUCTION

A running engineering machine is a source of mechanical vibrations with a wide spectrum of frequencies, including low frequencies [1], [2] and [3]. The vibrations act on the operator inside the machine [4], on all the machine subassemblies and subsystems and indirectly, on the surrounding environment. For the sake of the health of the machine's valves, it is essential to identify the mechanical vibrations to which they are subjected. Such vibrations often may disturb the operation of the entire hydraulic system of a mobile machine. A disturbance in the operation of such a system is reflected in a change in the pressure fluctuation spectrum. The disturbance may lead to a deterioration in the accuracy of positioning the actuators, to uneven operation, shortening of the machine's life and sometimes to a higher level of low-frequency noise emitted [5]. Low-frequency vibrations and noise have a particularly adverse effect on hydraulic valves and the human being. In hydraulic valves they may excite the vibration of their control elements (such as the slide and the head) [6] and [7]. This occurs when the frequency of the external mechanical vibrations is close to that of the free vibrations of the valve control element. In the case of a human being, the vibrations via the skin mechanoreceptors transmit specific information to the central nervous system, causing reflex reactions of the human body [3], [8] and [9]. The vibrations are accompanied by noise [10], also with low-frequency components. The noise is the subject of EU standard regulations. Hydraulic equipment producers, however, rather seldom specify the operating requirements concerning the resistance of their products (e.g. valves) to mechanical vibrations. One of such rare examples is the proportional distribution valve Parker–Hannifin D1FP, whose product data sheet [11]specifies the vibration value (about 250 m/s^2) permissible with regard to its resistance to mechanical vibrations. The resistance of hydraulic valves to mechanical vibrations is usually tested in accordance with the relevant standards, such as, e.g. DIN-IEC68, part 2–6 [12] in the case of the D1FP distribution valve. Valve mechanical vibration resistance tests can also be conducted in accordance with other procedures described in Polish and international standards. Standard PN-IEC 68-2-59 1996 [13] specifies a method of testing electrotechnical subassemblies, equipment and other products (including hydraulic valves used in electrical applications) which during operation may be exposed to short-duration pulsating or oscillating forces generated by, e.g. seismic phenomena, an explosion or the vibrations of the machine in which they are installed. The tested product is excited by a certain number of constant-frequency sinusoidal beats. Standard PN-EN 60068-2-6:2008 [14] specifies a method of testing subassemblies, equipment and other products, which during transport or operation may be exposed to harmonic vibrations generated mainly by rotating, pulsating or oscillating masses. Such excitations occur in ships, planes, terrestrial vehicles and space vehicles. Resistance to mechanical vibrations in a frequency range of 5–3000 Hz is tested. A critical frequency identified by the test is a frequency at which faulty operation of the product or a deterioration in its properties due to vibration manifests itself or mechanical resonances (e.g. a valve control element) occur. Polish standard PN-EN ISO 4413:2011 [15] includes, among other things, requirements for the assembly of hydraulic valves (also pumps, servomotors, filters, etc.), but limited to a general statement that one should consider the effect of gravitation and vibrations on the valve.

In general, the vibrational resistance standards specify the permissible level of external mechanical vibrations which may adversely affect a machine, a piece of equipment or their components. In the precision industry, the experimentally determined maximum acceleration of about $0.981 \ m/s^2$ has been adopted as the norm ensuring the vibrational resistance of measuring instruments and industrial equipment [16]. As regards rotor machines, they can be exposed to external mechanical vibrations below $9.81 \ m/s^2$, without any adverse effect on their operation.

In this paper, the possibility of reducing the vibrations of the hydraulic distribution valve is studied. The experimental studies have been broadened with a general theoretical analysis of the problem of reducing the vibrations of a valve or a control element, which can be helpful in selecting the characteristics of antivibration materials and vibration damping systems for this purpose.

EXPERIMENTAL STUDIES

Two ways of reducing the effect of external mechanical vibrations on hydraulic valve operation were investigated. One way consisted in flexibly fixing the distribution valve housing to a vibrating foundation while the other way consisted in introducing specially designed damping washers made of oil-resistant rubber into the distribution valve.

A linear hydrostatic drive simulator, capable of generating mechanical vibrations up to a frequency of 100 Hz, was used as the source of external vibrations. The simulator is described in more detail in [17]. As part of the experimental studies aimed at reducing distribution valve vibrations through flexible fixing, the acceleration of simulator table vibrations (excitation) and that of distribution valve housing vibrations (response) were measured (Fig. 1). In the case of the experiments aimed at reducing the vibrations of the distribution valve slide through the introduction of oil-resistant rubber washers into the distribution valve, the accelerations of both distribution valve vibrations (excitation) and slide vibrations (response) were measured (Fig. 2). The distribution valve was rigidly fixed in the simulator clamps.

Figure 1. Way of fixing distribution valve and arrangement of accelerometers: 1 – hydraulic distribution valve, 2 – simulator table clamp, 3 – set of elastomer washers, 4 – accelerometer for measuring simulator table vibration acceleration (excitation), 5 – accelerometer for measuring distribution valve housing vibration acceleration.

Figure 2. Way of fixing distribution valve and arrangement of accelerometers: 1 – simulator table clamp, 2 – hydraulic distribution valve, 3 – accelerometer for measuring distribution valve housing vibration acceleration (excitation), 4 – accelerometer for measuring distribution valve slide vibration acceleration (response).

PCB-ICP accelerometers, a signal conditioner VibAmp PA16000D, a Tektronix four-channel digital oscilloscope with dedicated software supplied by the producer and a PC for measurement data acquisition were used in the experiments.

Identification of Vibration IsolatorParameters

Prior to the experimental studies aimed at determining the possibilities of reducing the impact of external mechanical vibrations on the hydraulic distribution valve, the parameters (stiffness and damping) of the materials later used to reduce the vibrations of both the distribution valve housing and the slide were experimentally identified using cylindrical and cuboidal specimens and modal analysis equipment[18] and [19]. A load would be applied by means of a modal hammer incorporating a force gauge while the system response would be registered by an acceleration gauge. The measurement signals were recorded and processed by a dedicated analyser. An analysis of the signals was carried out using a PC with special software.

The aim of the tests was to determine the elasto-dissipative properties of the selected elastomer elements. The tests were carried out on a special test rig as shown in Fig. 3. The test rig consisted of:

- A seismic mass of 49 kg,
- A vibrating mass of 12.9 kg,
- Shock damping washers,
- And measuring equipment.

Figure 3. Test rig: 1 – tested specimens; 2 – vibrating mass; 3 – seismic mass, 4 – damping washers.

The measuring equipment consisted of a modal hammer (type SP 205) made by PCP Piezoelectronics, incorporating a force gauge for measuring the amplitude of the applied excitation, a PCB piezoelectric acceleration gauge for measuring the system response to the applied excitation and an HP analyser (type 35665A) recording the data from the gauges and converting them to transition characteristics.

The test rig was designed to minimize the influence of external vibrations on the motion of the vibrating mass. For this purpose a seismic mass characterized by great stiffness and weight and isolated from the foundation by means of damping washers was used.

The aim of the tests was to determine the frequency characteristics for each of the types of specimens. The characteristics were used to calculate the damping and dynamic stiffness parameters. The specimens would be placed on the seismic mass and a steel element constituting the vibrating mass would be placed on them. All the specimens of the same type would be arranged in the same precisely specified way. Because of the small size of the cylindrical specimens, four such specimens would be tested simultaneously in order to determine their elasto-dissipative properties. The specimens would be arranged to form a square with a side of 14 cm. This ensured the uniform loading of the specimens during the dynamic tests. The cuboidal specimens would be arranged in pairs parallel to one another so that the vibrating mass oscillated along the vertical axis.

In the case of the cylindrical elastomer washers, measurements would be conducted for two arrangements of the specimens. In the first arrangement the long axis (symmetry axis) of the specimen was oriented along the vertical axis, i.e. parallel to the direction of motion of the vibrating mass. In the second arrangement, the long axis of the specimen lay in the horizontal plane, i.e. it was oriented perpendicularly to the direction of vibrations. The specimens and their designations are shown in Fig. 4. The letter W represents a cylindrical specimen while the letter C or Z (in the second position) stands for the kind of material: Adipol 70 ShA and Ultraflex 64 ShA, respectively. The letters T and L indicate the direction of specimen loading: L – along the specimen's axis of symmetry, T – transversely to the symmetry axis. Digit 1 at the end stands for a 25 mm long specimen while digit 2 indicates a specimen length of 16 mm. For all the cylindrical elastomers the outside diameter was 16 mm and the inside diameter amounted to 6 mm.

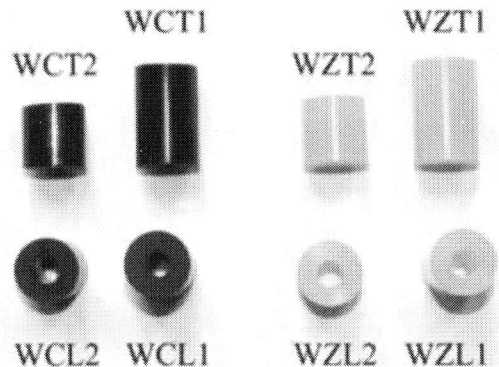

Figure 4. Designations of cylindrical specimens – top view.

Tests were also carried out on cuboidal specimens made of oil-resistant rubber that later would be placed in the clamps for reducing distribution valve vibrations. All the specimens had the same length and width, i.e. 101 and 26 mm, respectively, whereas their thickness amounted to 4, 12 and 15 mm. Their designations include digits indicating specimen thickness while the symbol TR represents a material with enhanced stiffness. The identification tests carried out on the test rig described above and shown in Fig. 3yielded force–time and acceleration–time diagrams which were used to determine the transition characteristic by means of the HP 3566 analyser. An exemplary experimental transition characteristic for the specimen designated with symbol 12 is shown in Fig. 5.

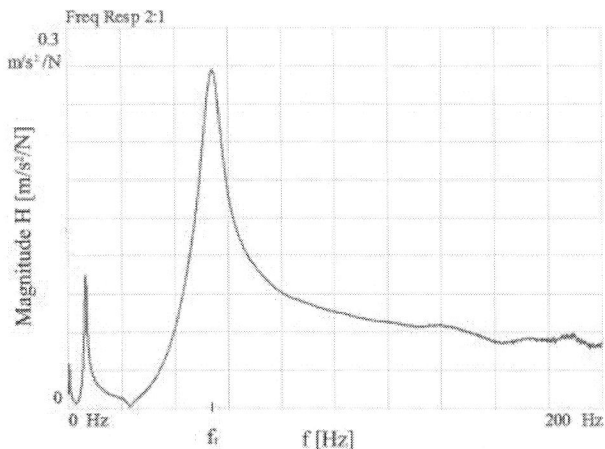

Figure 5. Frequency characteristic for cuboidal specimen 12 made of oil-resistant rubber; f – vibration frequency of vibrating mass (item 2,Fig. 3), f_r – vibrating mass natural frequency.

From the frequency characteristics the natural frequency of the elastomer-vibrating mass system and the corresponding modal damping were determined. Then the damping and dynamic stiffness of the tested material were calculated.

The modal damping was calculated, using the half-power method [19], from the formula:

$$\xi_r = \frac{\omega_2 - \omega_1}{2\omega_r} \tag{1}$$

where ω_1 and ω_2 were calculated from the relation:

$$|H(j\omega_1)| = |H(j\omega_2)| = \frac{|H(j\omega_r)|}{\sqrt{2}} \tag{2}$$

where $H(j\omega r)$ – an amplitude value [m/s²/N] corresponding to natural angular frequency ωr, $H(j\omega_1)$, $H(j\omega_2)$ – amplitudes [m/s²/N] corresponding to angular frequency ω_1 and ω_2, lying on both sides of natural angular frequency ωr, consistently with the condition: (2), $\omega r = 2\pi fr$, fr – natural frequency.

In order to describe the dynamic properties of the tested materials one can use a dynamic model with one degree of freedom (Fig. 6) [19], consisting of linear elasto-dissipative elements connected in parallel (a Kelvin–Voigt body) where mass m corresponds to the vibrating mass used in the test.

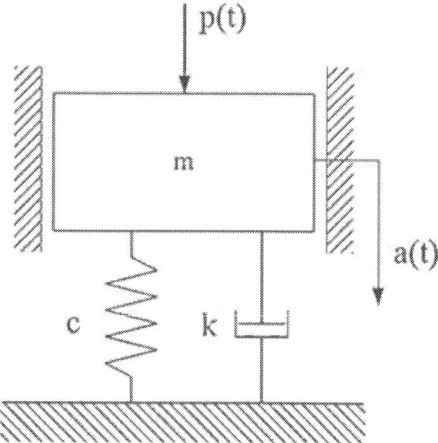

Figure 6. Schematic of dynamic model: c – stiffness parameter, k – damping parameter, m – vibrating mass.

Considering that the tested elements were so arranged that they could be treated as connected in parallel, the unit stiffness and damping for the cylindrical specimens can be calculated from the formulas:

$$c = \frac{c_z}{4}; \quad k = \frac{k_z}{4} \tag{3}$$

where kz and cz are respectively the equivalent damping value and the equivalent stiffness value for the set of tested specimens, and for the cuboidal specimens from:

$$c = \frac{c_z}{2}; \quad k = \frac{k_z}{2} \tag{4}$$

The calculated parameters are presented in Table 1 and Table 2.

Table 1. Damping k and stiffness c parameters for cylindrical specimens.

	Damping k [kg/s]	Stiffness c [kg/s^2]
WCL1	10.97	9.70E+04
WCT1	7.26	4.30E+04
WCL2	9.27	8.04E+04
WCT2	8.47	6.49E+04
WZL1	7.01	8.41E+04
WZT1	3.55	2.79E+04
WZL2	4.69	5.55E+04
WZT2	4.43	4.30E+04

Table 2. Damping k and stiffness c parameters for cuboidal specimens.

	Damping k [kg/s]	Stiffness c [kg/s^2]
4	112.88	2.16E+06
12	59.02	7.50E+05
15	52.57	7.16E+05
TR4	149.96	4.08E+06
TR12	91.27	1.44E+06
TR15	66.11	1.16E+06

The stiffness c and damping k parameter values experimentally determined for the materials were then used in studies aimed at reducing the impact of external vibrations on the distribution valve and the slide. The obtained results also show, for all the types and dimensions of the elastomers, that when the long axis of the specimen was oriented parallel to the direction of motion of the vibrating mass, the tested elements were characterized by greater damping and stiffness than when the long axis was oriented perpendicularly to the direction of load application. Moreover, the elastomers made of Ultraflex 64 ShA are characterized by much higher damping and stiffness values than the elastomers made of Adipol 70 ShA. In the case of the cuboidal specimens, damping and stiffness were found to decrease with increasing specimen thickness.

Reduction of Distribution Valve Housing Vibrations

A series of experiments were carried out in which the distribution valve would be fixed in the simulator table clamps, using sets of antivibration insulation elements (whose elasto-dissipative properties had been experimentally identified – Table 1 and Table 2) in different configurations (Fig. 7 and Fig. 8). The distribution valve would be excited with harmonic simulator table vibrations with a frequency of 10–100 Hz and a known amplitude.

Figure 7. Hydraulic distribution valve fixed in clamps with flexible antivibration insulation elements: 1 – simulator table, 2 – clamp, 3 – hydraulic distribution valve, 4 – set of flexible antivibration insulation elements.

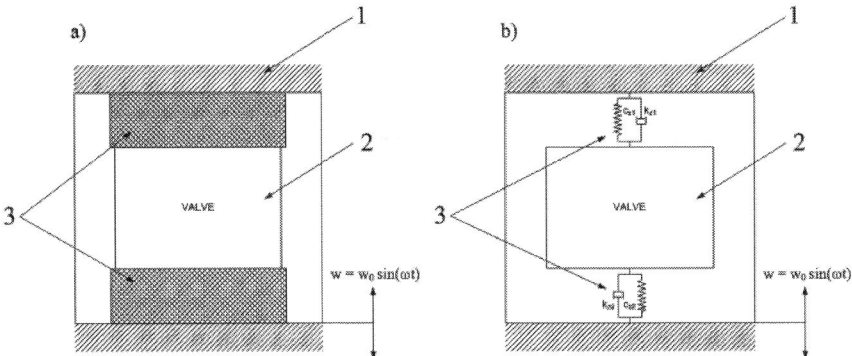

Figure 8. Schematic of flexible fixing of distribution valve to foundation: (a) location of flexible elements, (b) dynamic model of flexible elements: 1 – vibrating simulator table clamp, 2 – hydraulic distribution valve, 3 – set of flexible antivibration insulation elements.

Different sets of elastomer damping washers would be placed on both sides of the distribution valve in the clamps. In order to assess the effectiveness of the antivibration insulation provided by the elastomer elements the ratio of the distribution valve vibration acceleration amplitude (response) to the simulator table vibration acceleration amplitude (excitation) was determined. It was assumed that antivibration insulation was effective when the following inequality was satisfied:

$$\frac{a_2}{a_0} < 1$$

(5)

where a_2 – the amplitude of distribution valve housing vibration acceleration [m/s²], a_0 – the amplitude of simulator table vibration acceleration [m/s²].

Five different sets of vibration isolators were used in the tests. The particular sets differed from each other in the number of elastomer elements which they consisted of, and, for the cylindrical elements, in the direction of load application relative to the long axis. Hence the sets were characterized by different equivalent stiffness cz values and different equivalent damping kz values, as shown in Table 3.

Table 3. Equivalent stiffness cz and equivalent damping kz values for selected sets of washers.

Washer set number	Equivalent stiffness cz [N/m]	Equivalent damping kz [kg/s]
1	2.84E+05	43.24
2	2.32E+06	132.22
3	5.44E+05	55.84
4	6.55E+05	70.04
5	7.16E+05	84.88

The experimental results in the form of a diagram showing the distribution valve acceleration amplitude/simulator table vibration acceleration amplitude ratio depending on the excitation frequency are presented in Fig. 9.

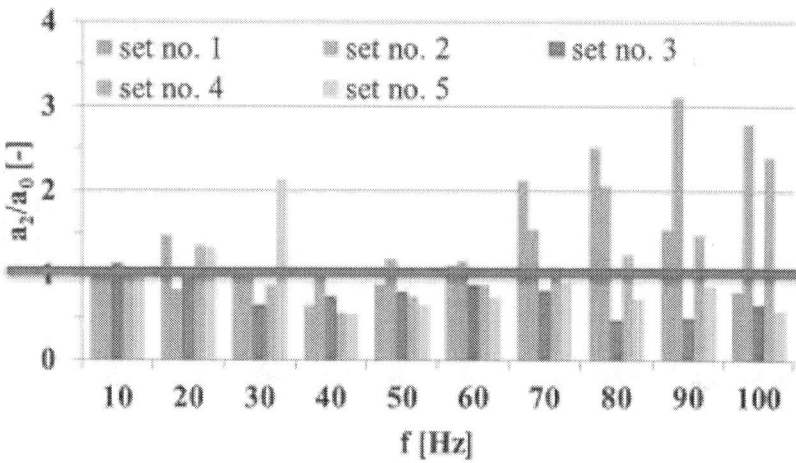

Figure 9. Effectiveness of damping valve vibrations excited by foundation vibrations for different sets of flexible washers: a_2 – valve housing vibration acceleration amplitude, a_0 – simulator table vibration acceleration amplitude.

It appears from the results presented in Fig. 9 that such a set of washers can be designed which will ensure effective antivibration insulation in nearly the whole considered range of frequencies.

Reduction of Slide Vibrations
In order to explore the possibilities of reducing slide vibrations, experiments were carried out in which specially shaped washers made of

oil-resistant rubber would be installed inside the distribution valve, i.e. between the valve housing and the slide centring springs. The distribution valve would be fixed directly in the clamps of the simulator table. The simulator table vibrations ranged from 10 Hz to 100 Hz. The arrangement of the washers inside the distribution valve housing and their shape are shown in Fig. 10.

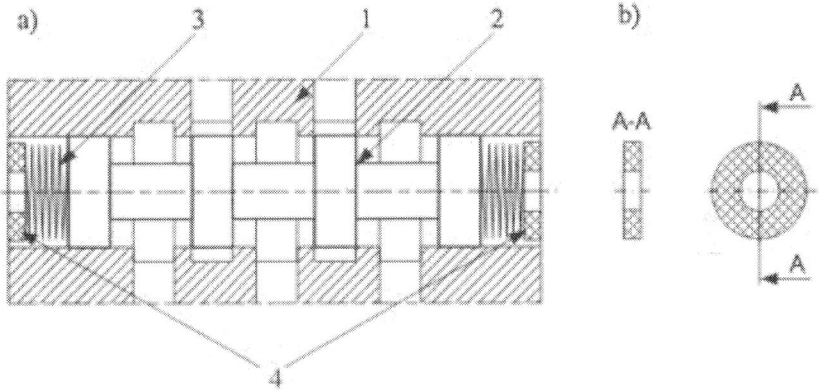

Figure 10. Oil-resistant rubber washers installed inside distribution valve housing: (a) location inside valve housing: 1 – vibrating valve housing, 2 – valve slide, 3 – centring springs, 4 – flexible washers, (b) washer shape.

Because of the distribution valve design, washers with an outside diameter of 26 mm, an inside diameter of 22 mm and a height of 4 mm were used. The dynamic model of the slide inside the valve housing is shown in Fig. 11. The slide having mass m_1 is centred by springs each with known stiffness cs_1. Moreover, damping ks_1 occurs on both sides of the slide.

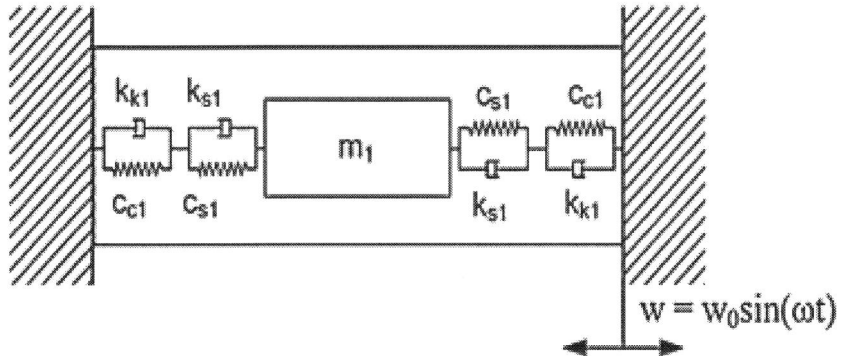

Figure 11. Dynamic model of slide in vibrating distribution valve housing.

The washers are represented by a model consisting of linear elasto-dissipative elements connected in parallel (the Kelvin–Voigt model), with one degree of freedom, where stiffness is denoted with cs_1 and damping with kk_1, for each of the two washers.

The centring spring stiffness in the tested distribution valve was $cs_1 = 2900$ N/m. The washer stiffness was $cc_1 = 1.41E+06$ N/m. Therefore, assuming the connection to be linear, the equivalent stiffness on both sides of the slide amounts to 2894 N/m. Where as the equivalent stiffness on both sides of the slide for the parallel connection amounts to 5788 N/m. In a similar way the equivalent damping for one side was found to amount to 8.66 kg/s while the equivalent damping on both sides of the slide for the parallel connection amounted to 17.32 kg/s. The experimental results showing the effect of the use of the washers in reducing the slide vibrations are presented in Fig. 12.

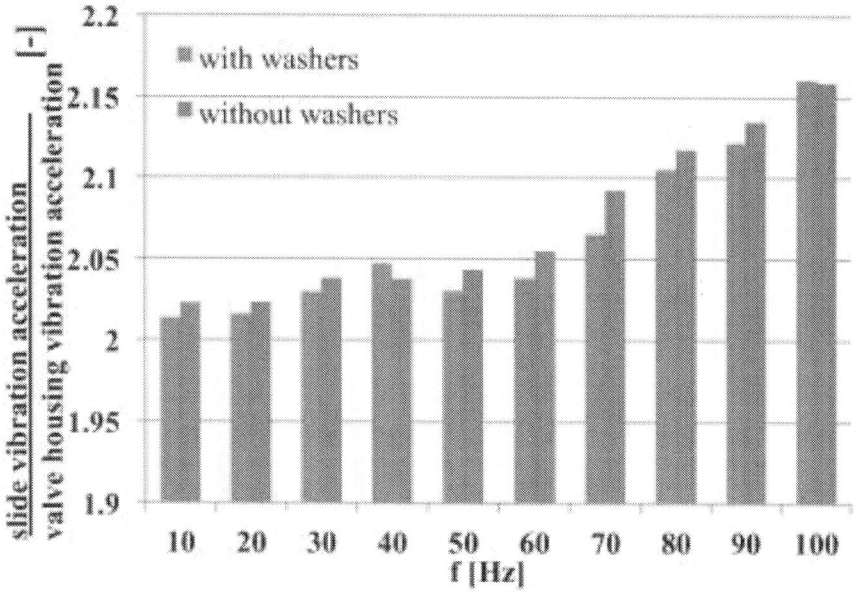

Figure 12. Relative slide vibration acceleration/valve housing vibration acceleration ratio versus valve housing vibration frequency.

An analysis of the equivalent stiffness on each side of the slide indicates that it is decisively influenced by the lower of the values, i.e. the spring stiffness in this case. This is due to the way in which the stiffness is connected, i.e. in series. The results presented in Fig. 12 show that the adopted solution does not result in a significant reduction of slide vibrations. Therefore the effect of a change in stiffness and damping on the amplitude of the vibration of the slide excited into vibration by the vibrations of the valve housing should be examined. In the case of distribution valves controlled by proportional electromagnets, one should take into account the maximum forces (typically 15–20 N [20] and [21]) generated by such electromagnets with an adjusted step, and the size of the latter (typically about 4.5 mm [20] and [21]) required by the slide pair design. This puts constraints on the washer stiffness value since the controlling force generated by the electromagnet must be greater than the sum of the other forces, i.e. the friction force in the slide pair, the inertial force of the slide and the associated fluid, the equivalent stiffness force of the centring springs and washers and the hydrodynamic force.

THEORETICAL ANALYSIS OF VIBRATION REDUCTION POSSIBILITIES

The experimental studies indicate that when selecting a way of reducing the impact of external mechanical vibrations on the hydraulic distribution valve and its slide one should take into consideration the constraints relating to the dimensions of vibration isolators and their stiffness c and damping k. Geometric constraints stem from the incorporation conditions, i.e. the space available for installing a vibration isolator (e.g. a set of elastomer washers). Isolator stiffness c constraints stem from the maximum power generated by the slide controlling element (e.g. a proportional electromagnet).

The theoretical analysis can be based on the same mathematical model used in two ways: one can consider a reduction in valve housing vibration as an excitation acting on the valve slide or a reduction in the relative vibrations of the slide inside the valve housing. In both cases one can use a single mass model with one degree of freedom, in which m represents, depending on the considered case, the distribution valve mass or the slide mass, where c and k are respectively the equivalent stiffness and the equivalent damping of the flexible washers between the distribution valve housing and the clamps or the equivalent stiffness of the flexible washers and the centring springs and the equivalent damping of the flexible washers of the slide pair (Fig. 13).

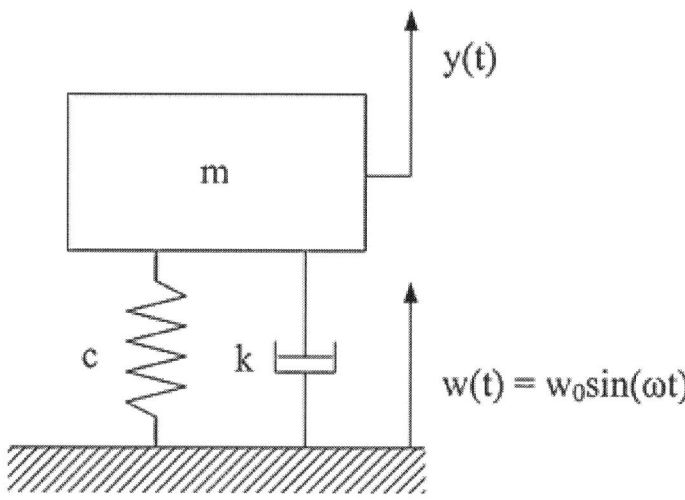

Figure 13. Model of vibrating system with one degree of freedom.

A body having mass m is excited into vibration by a kinematic excitation in the form of a harmonic function expressed by the equation:

$$w = w_0 \sin(\omega t) \tag{6}$$

where w_0 – vibration amplitude [m], $\omega = 2\pi f t$ [rad/s], f – frequency [Hz], t – time [s].

The absolute motion of the body (the distribution valve housing or the slide) is described by the equation:

$$m \cdot \ddot{x} + k(\dot{x} - \dot{w}) + c(x - w) = 0 \tag{7}$$

The relative displacement of the body with mass m and the vibrating foundation (e.g. a distribution valve housing) can be written as:

$$y = x - w \tag{8}$$

Thus the relative motion equation assumes the form:

$$m\ddot{y} + m\ddot{w} + k\dot{y} + cy = 0 \tag{9}$$

Considering that w is a known time function, Eq. (9) can be written as:

$$\ddot{y} + 2h\dot{y} + \omega_0^2 y = -\ddot{w}(t) \tag{10}$$

And after the harmonic form of the kinematic excitation is taken into account, as:

$$\ddot{y} + 2h\dot{y} + \omega_0^2 y = w_0\omega^2 \sin(\omega t) \tag{11}$$

where

$$h = \frac{k}{2m} \tag{12}$$

$$\omega_0^2 = \frac{c}{m} \tag{13}$$

The relative vibrations of the distribution valve slide are described by the equation:

$$y(t) = Baw_0\sin(\omega t + \delta) = Baw(t + \tau) \tag{14}$$

The slide vibrations are proportional to the housing vibrations, but they are shifted by time $\tau = \delta/\omega$. Coefficient Ba, which can be called a transfer factor, amounts to [22]:

$$B_a = \frac{(\omega/\omega_0)^2}{\sqrt{(1-(\omega^2/\omega_0^2))^2 + 4(h^2/\omega_0^2)(\omega^2/\omega_0^2)}} \tag{15}$$

Or

$$B_a = \frac{y_0}{w_0} \tag{16}$$

where y_0 – the amplitude of relative slider displacement, w_0 – the amplitude of valve housing vibrations.

For given kinematic excitation parameters one can plot a spatial diagram of the dependence between transfer factor Ba and the equivalent stiffness c (of the centring springs and the flexible washers) and the equivalent damping k inside the distribution valve housing. Fig. 14 shows such a dependence for the assumed parameters: slider mass $m = 0.18$ kg and excitation frequency $f = 80$ Hz.

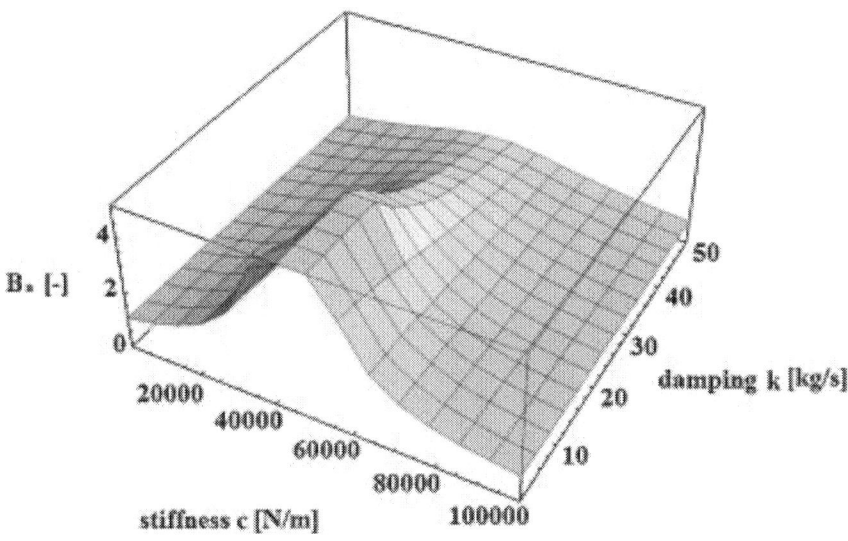

Figure 14. Dependence between transfer factor B_a and equivalent stiffness c and damping k inside distribution valve.

In order to reduce relative slide vibrations one should select such an equivalent stiffness c and/or an equivalent damping k inside the distribution valve housing that $Ba < 1$. Assuming an equivalent damping k inside the valve housing equal to 25 kg/s and $f = 80$ Hz, one can determine the equivalent stiffness c value for which $Ba < 1$ (Fig. 15).

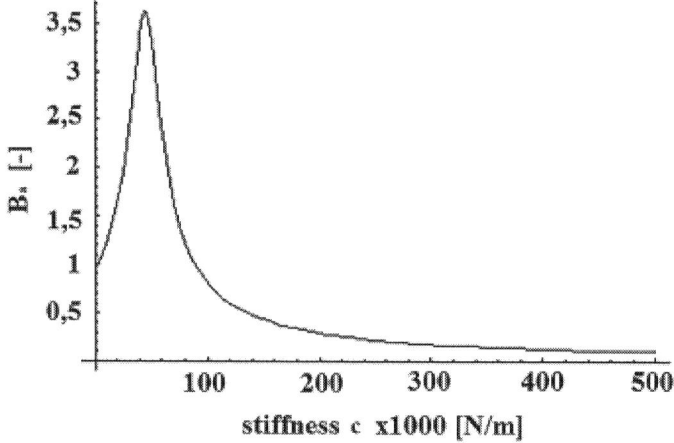

Figure 15. Factor B_a as function of equivalent stiffness c for adopted damping k and excitation parameters.

It follows from Fig. 15 that equivalent stiffness c should be higher than 100,000 N/m in order to avoid slider resonance for the adopted excitation parameters. However, considering the controlling forces, this is not possible to achieve because of the forces (typically about 15 N) generated by the proportional electromagnets. Assuming a slide stroke of 4.5 mm, one gets the maximum equivalent stiffness of 3333 N/m (excluding the other forces which the controlling force generated by the electromagnets must overcome).

If the equivalent stiffness c equal to 3000 N/m and the same excitation parameters as above are assumed, one can examine the influence of equivalent damping k on factor Ba (Fig. 16).

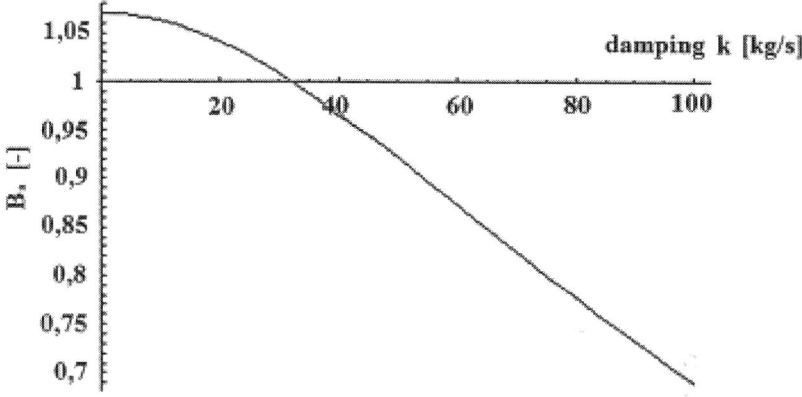

Figure 16. Factor B_a as function of equivalent damping k for adopted stiffness c and excitation parameters.

According to Fig. 16, the equivalent damping k for the adopted equivalent stiffness c should be greater than 30 kg/s.

A similar analysis, aimed at minimizing the amplitude of absolute housing vibrations, can be carried out for the distribution valve housing [23]:

$$x_0 = w_0 \cdot \sqrt{\frac{1+(2\gamma(\omega/\omega_0))^2}{(1-(\omega/\omega_0)^2)^2 + (2\gamma(\omega/\omega_0))^2}} \rightarrow \min \qquad (17)$$

where w_0 – the excitation amplitude [m], ω – the excitation angular frequency [rad/s], ω_0 – the system natural angular frequency [rad/s], γ – a dimensionless damping coefficient, $\gamma=(h/\omega_0)=(k/2m\omega_0)$.

In addition, one can use the antivibration insulation effectiveness calculated from the relation [3]:

$$\varepsilon = \left(1 - \frac{x_0}{w_0}\right) \times 100\% \tag{18}$$

In this case, such parameters c and k should be selected that the antivibration insulation effectiveness, calculated from formula (18), is as high as possible.

CONCLUSIONS

The results of the experiments and the theoretical analysis indicate that in order to reduce slider vibrations one can introduce shock damping washers (made of a material characterized by high stiffness c and damping k) into the distribution valve housing, between the housing and the centring springs. In the case of distribution valves with single-step electric (e.g. proportional) control, this approach has constraints because of the maximum values of the controlling forces generated by the proportional electromagnets, and the required slide stroke. Another possible way of reducing distribution valve housing vibrations, and consequently of slider vibrations, is to mount the distribution valve on flexible washers whose equivalent stiffness and equivalent damping can be calculated from relation (17) for the set excitation parameters and the mass of the vibrating valve. The experimental results presented in Fig. 9 and Fig. 12 indicate that although the materials used reduce slider or housing vibrations, the antivibration insulation effectiveness is not satisfactory. Therefore, mainly for the purposes of isolating distribution valve housing vibrations, one should search for other materials, using the criteria defined by relations (17) and (18).

On the basis of the theoretical analysis the following simplified and generalized procedure for selecting antivibration insulation is proposed [3]:

- Determine the required antivibration insulation effectiveness; usually the effectiveness level of 70% is sufficient in engineering practice.
- For the antivibration insulation effectiveness determined in pt. 1, specify the maximum value of ratio x_0/w_0, using Table 4 [3].

Table 4. Effectiveness of antivibration insulation [3].

Antivibration insulation effectiveness [%]	Maximum antivibration insulation factor	Required ratio ω/ω_0
90	0.1	3.32
80	0.2	2.45
70	0.3	2.08
60	0.4	1.87
50	0.5	1.73

- Determine the value of the ratio of excitation angular frequency ω to distribution valve natural angular frequency ω_0. For this purpose determine the lowest value of excitation angular frequency ω.
- From the nomograph used for determining antivibration insulation parameters select natural angular frequency ω_0 of the distribution valve to be insulated, which is needed to ensure the antivibration insulation factor value determined in pt. 2. The nomograph can be found in the literature, e.g.[3] and [24].
- From the nomograph determine the static deflection corresponding the natural angular frequency ω_0 specified in pt. 4.
- Calculate equivalent stiffness c, from relation (13).
- Calculate the stiffness of the particular insulating elements, using relation $ci = c/n$, where n – the number of insulating elements;
- Calculate the load per insulating element.
- Select an antivibration insulation system from the typical commercial antivibration insulation products or design your own system satisfying the above requirements.

REFERENCES

1. J. Sadowski, R. Rogiński, Noise Control in Public Transport, WKŁ, Warsaw, 1965 (in Polish).
2. J. Grajnert, Vibration Insulation of Machines and Vehicles, Wrocław University of Technology Publishing House, Wrocław, 1997 (in Polish).
3. Z. Engel, M. Zawieska, Noise and Vibration in Work Processes: Sources, Assessment, Hazards, Central Institute for Labour Protection – National Research Institute, Warsaw, 2010 (in Polish).
4. M. Demic, J. Lukic, Investigation of the transmission of fore and aft vibration through the human body, Applied Ergonomics 40 (2009) 622–629.

5. N. Chenxiao, Z. Xushe, Study on vibration and noise for the hydraulic system of hydraulic hoist, in: Proceedings of 2012 International Conference on Mechanical Engineering and Material Science (MEMS), 2012.

6. M. Stosiak, The effect of ground mechanical vibrations on pressure fluctuation in the hydraulic system, HydraulikaiPneumatyka 26 (3) (2006) 5–8 (in Polish).

7. M. Stosiak, Vibration insulation of hydraulic system control components, Archives of Civil and Mechanical Engineering 11 (1) (2011) 237–248.

8. J. Renowski, Noise: Indices and Evaluation Criteria, Wrocław University of Technology Publishing House, Wrocław, 1988 (in Polish).

9. Z. Kudźma, M. Stosiak, Reduction of infrasounds in machines with hydrostatic drive, Acta of Bioengineering and Biomechanics 15 (2) (2013) 51–64.

10. H. Ortwig, Experimental and analytical vibration analysis in fluid power systems, International Journal of Solids and Structures 42 (2005) 5821–5830.

11. D1FP – Parker–Hannifin valve data sheet.

12. DIN IEC 68 Part 2–6, Fundamental environmental test procedures, sinusoidal oscillation.

13. PN-IEC 68-2-59:1996, Environmental research. Fe test: vibration – sinusoidal beat method (in Polish).

14. PN-EN 60068-2-6:2008, Environmental research – part 2–6: tests – Fe test: vibration (in Polish).

15. PN-EN ISO 4413:2011, Hydraulic drives and controls – general guidelines for systems (in Polish).

16. R. Łączkowski, Vibroacoustics of Machines and Equipment, WNT, Warsaw, 1983 (in Polish).

17. W. Kollek, Z. Kudźma, M. Stosiak, Simulator of linear hydrostatic drive as source of new research possibilities, in: Proceedings of Hydraulic and Pneumatic Drives and Controls 2005, Problems and Development Trends in the First Decade of the 21st Century, International Scientific-Technical Conference, 2005.

18. T. Kucharski, System of Measuring Mechanical Vibrations, WNT, Warsaw, 2002 (in Polish).

19. M. Kulisiewicz, S. Piesiak, Methodology for Modelling and Identifying Dynamic Mechanical Systems, Wrocław University of Technology Publishing House, Wrocław, 1994 (in Polish).

20. E. Tomasiak, Hydraulic and Pneumatic Drives and Controls, Silesian Polytechnic Publishing House, Gliwice, 2001(in Polish).

21. Osiecki, Hydrostatic Drive for Machines, WNT, Warsaw, 1998 (in Polish).

22. Z. Osiński, Theory of Vibrations, PWN, Warsaw, 1980 (in Polish).

23. J. Goliński, Vibro-Insulation of Machines and Equipment, WNT, Warsaw, 1979 (in Polish).

24. C.M. Harris, A.G. Piersol, Shock and Vibration Handbook, McGraw-Hill, New York, 2002.

CITATION

M. Stosiak, Ways of reducing the impact of mechanical vibrations on hydraulic valves, Archives of Civil and Mechanical Engineering, Volume 15, Issue 2, February 2015, Pages 392-400, ISSN 1644-9665, http://dx.doi.org/10.1016/j.acme.2014.06.003.

CHAPTER 3

Active Structural Acoustic Control of Repetitive Impact Noise

G. Pinte, R. Boonen, W. Desmet, P. Sas

Department of Mechanical Engineering, KatholiekeUniversiteit Leuven, Celestijnenlaan 300B, B-3001 Heverlee, Belgium

ABSTRACT

This paper discusses the effectiveness of an Active Structural Acoustic Control (ASAC) system for the reduction of repetitive impact noise, radiated by structures with a high modal density in the controlled frequency range. Although there is a significant difference in nature between periodic and transient noise, up till now no specific research on ASAC of transient noise was reported. The development of the ASAC system is divided into two phases: the definition of the control configuration and the design of a suitable control algorithm. The optimal control configuration as well as the implemented control algorithm for the reduction of impact noise differ significantly from the common solutions in periodic noise control. In the first part of the paper, a practical methodology is presented to define a good control arrangement for transient noise control. The second part of the paper focuses on the design of control algorithms, adapted to the specific properties of impact noise. Since many industrial impact noise problems involve successive impacts with a repetitive behaviour, control algorithms with a learning behaviour are discussed. The efficiency of these Iterative Learning Control (ILC) algorithms is extensively demonstrated in this paper. The developed ASAC strategy has been verified on a thick steel plate, which is excited by successive impacts. The obtained results show that ASAC can be a very efficient transient noise control technique in certain industrial applications (e.g. presses, punching machines, etc.).

INTRODUCTION

Noise pollution from modern industrial activities is an environmental problem of growing importance. Especially in machine halls with production machines such as punching machines and presses, that generate impact noise, the radiated noise levels exceed often the legal regulations regarding human exposure to noise [1] and [2]. The noise, radiated by these machines, is mainly structure-borne, i.e. generated by structural vibrations of different mechanical parts. Various studies [3] and [4] have shown that Active Structural Acoustic Control (ASAC) has some potential to reduce stationary structure-borne noise. In ASAC the sound field radiated by a structure is controlled by intervening in the structural vibrations with actuators. Although there is a significant difference between periodic and transient noise, up till now almost no particular research on ASAC of transient noise has been reported, at least not to the authors' knowledge. The goal of the presented research is to study the possibilities of ASAC techniques to reduce transient structure-borne noise. The efficiency of an ASAC system depends on the characteristics of the control actuators and sensors, on their configuration as well as on the implemented control algorithm. First of all, for the reduction of transient noise, actuators should be available, which can deliver the necessary energy levels within the required (short) time intervals. The physical arrangement of the control system defines the maximum achievable performance, while the controller design determines how close the practical system approximates this performance.

The first part of the paper describes a methodology to define the optimal control configuration. Based on experimental and simulation results, the best configuration for an optimal global performance of the practical ASAC system can be found.

In the second part of the paper, the algorithm for the control of the actuator(s) is designed. The existing ASAC control algorithms for the reduction of stationary noise can be classified into two groups: feedforward and feedback control algorithms. The standard adaptive feedforward algorithms [5] (e.g. filtered-x LMS algorithm), which are nowadays most popular in ASAC applications, cannot be used for the control of transient noise signals, since the adaptive part is often based on a continuous convergence to an optimum. Recently some new feedforward algorithms with specific adaptive filters [6] are proposed for the control of impulsive noise. Contrary to the conventional feedforward algorithms, the commonly used feedback algorithms developed for the control of continuous noise can be used for transient noise. However, the use of these feedback controllers is limited for practical, industrial applications: due to the high modal density of the controlled structures in the frequency range of interest, it is impossible to design efficient feedback controllers of limited order. Since in many industrial impact noise problems the successive impacts have a repetitive behaviour (e.g. a punching machine often performs the same operations several times), a learning behaviour can be introduced in the control algorithms, resulting in an improved performance of the controller as the number of controlled impacts increases. This paper presents the development of Iterative Learning Control (ILC) algorithms, which can be considered as adaptive feedforward algorithms. In an ILC controller the control signal is adapted to the specific repetitive properties of the transient noise from successive impacts. Because in industrial machinery, which radiates impact noise, advanced knowledge is often available of when the impact will occur, the ILC control filters can be noncausal: in this way the ILC controller can anticipate for a future impact. The advantages and drawbacks of causal and noncausal filtering in ILC algorithms are discussed in this paper. In the final ASAC controller, a combination of an ILC and a feedback algorithm is implemented, which is an example of a hybrid feedforward–feedback system as proposed in Refs. [7] and [8].

The last part of the paper presents the global noise reduction results, achieved by the developed ASAC systems. Several control configurations and algorithms are compared and the performance and the practical limitations of various strategies are discussed.

DEMONSTRATOR

In this paper, the developed ASAC strategies are used to cancel the low frequency $(< 1000\,\text{Hz})$ noise radiated by a thick free–free suspended plate $(500\,\text{mm} \times 600\,\text{mm} \times 15\,\text{mm})$, which is excited by successive, repetitive impacts (Fig. 1). This set-up is studied, because its dynamic and acoustic behaviour resembles that of the massive frames of production machines, which radiate impact noise (e.g. punching machines, presses): the plate, which has a high modal density in the controlled frequency range below 1000 Hz (Table 1), radiates structureborne noise. A plate is an ideal demonstration breadboard since it is easy to define a reliable model to evaluate the theoretical possibilities of different ASAC controllers.

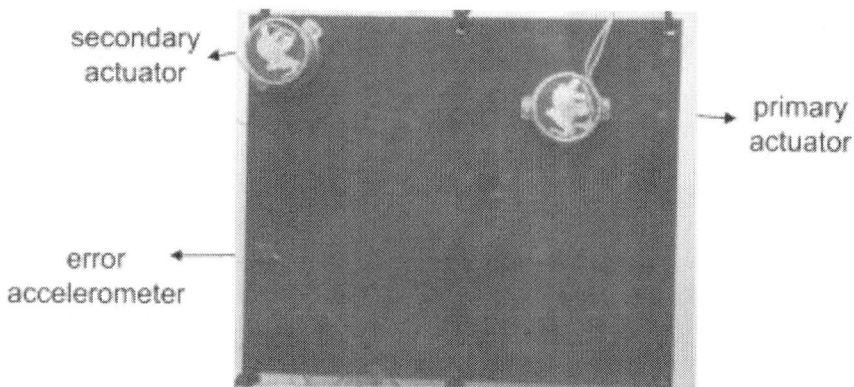

secondary actuator

primary actuator

error accelerometer

Figure 1. The impact plate test setup with a primary disturbance and secondary control shaker and an error accelerometer.

Table 1. The resonance frequencies of the plate below 1000 Hz, defined by an experimental identification

Resonance number	1	2	3	4	5	6	7	8	9	10	11
Resonance frequency (Hz)	160	205	227	325	399	458	611	772	796	881	991

The disturbing impact forces on the plate are not caused by the actual collision of an object and the plate, but are simulated by a primary inertia shaker, generating force pulses on the plate with a duration of 2 ms. In this way, especially the low-frequency modes of the plate below 1000 Hz are excited, which is the frequency range to tackle by active means. Noise, which is radiated by the plate at higher frequencies, can be damped more efficiently by passive control techniques.

Since the successive input signals to the primary actuator are identical, the impact forces on the plate as well as the resulting vibrations and the radiated noise pulses are repetitive. The time between two consecutive impacts is variable but is always longer than 4 s such that the vibrations, due to a previous impact, are totally damped out when a new impact is generated. Consequently, the control of a new impact is not influenced by the previous impacts. Analogously to industrial machinery, where advanced knowledge is often available about the moment of the coming impact, in the demonstrator a trigger signal becomes active 0.5 s before a new voltage pulse is sent to the primary shaker.

Fig. 2(a) shows the A-weighted radiated noise levels in the one-third octave bands below 1000 Hz, which are measured in the far field when the plate is excited by the primary shaker. Most of the radiated acoustic energy is situated in the one-third octave bands with centre frequencies 630 and 800 Hz. A detailed autopowerspectrum of the radiated noise in this frequency range from 500 to 900 Hz is given in Fig. 2(b): the spectrum is mainly dominated by three efficiently radiating resonance frequencies of the plate (at 611, 772 and 881 Hz). Consequently the ASAC strategy should focus on noise reduction of the corresponding structural modes between 500 and 900 Hz.

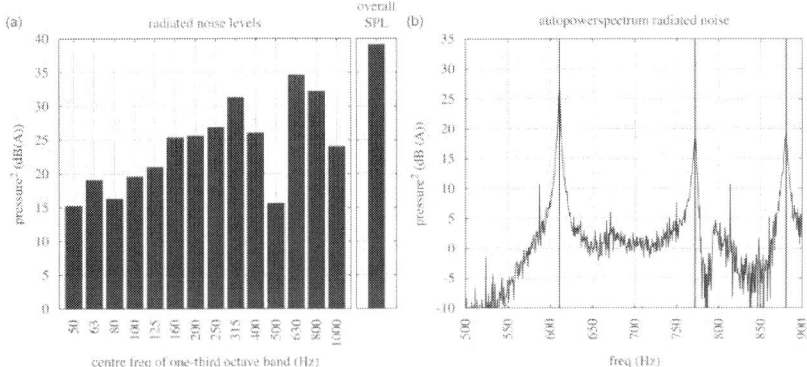

Figure 2. The radiated noise, measured in the far field when the plate was excited by the primary shaker: (a) A-weighted noise levels in the one-third octave bands below 1000 Hz and A-weighted overall Sound Power Level and (b) detailed A-weighted autopowerspectrum between 500 and 900 Hz.

CONTROL CONFIGURATION

The first phase in the development of an ASAC system is the definition of the optimal control configuration, i.e. the choice of the number and the location of actuator(s) and sensor(s). Many studies, both numerical and experimental, have already been devoted to this research topic. Two approaches can be distinguished, based on the goal function which is considered: some researchers minimize the vibration or acoustic level at certain locations, while others use vibration or acoustic power as cost function [9]. Simple problems (ASAC of beams [10], plates [11] and [12], etc.) can be described analytically such that different control configurations can be compared quite easily and some general design guidelines can be developed. Because the dynamic and acoustic behaviour of more complicated structures can only be evaluated by numerical models (e.g. finite element models, boundary element models, etc.) or extensive experimental models, the definition of the optimal control configuration demands computationally intensive optimisation algorithms (e.g. genetic algorithms [13] and [14]). Therefore, in many practical ASAC applications, the measuring/modelling and optimisation effort will be too high, even for a limited number of actuators and sensors. Consequently, simpler design rules, which can find sub-optimal but satisfactory configurations, are required. In this paragraph, a new methodology is presented to find suitable locations for actuators and sensors in the case of ASAC of impact noise. This methodology is a 2-step procedure, which first defines the actuator locations and afterwards searches compatible sensor locations.

Previous research [15] shows that, if a limited number of actuators and sensors is used, the optimal control arrangement is strongly frequency dependent. Since in the case of an impact excitation a broadband frequency range is excited, the performance of the control system should be balanced over this whole frequency range. In the excited frequency range, the structure will mainly vibrate and radiate noise at certain efficiently radiating resonance frequencies such that an efficient vibration reduction at these frequencies mainly determines the overall impact noise reduction and outweighs the control performance at other frequencies. Therefore, the actuators should be positioned at locations with a good controllability of the modes at these resonance frequencies. The controllability of different positions can be checked by calculating or measuring the resonance frequencies and corresponding mode shapes. The complexity of the structure and the impact duration, which defines the excited frequency range, determine the number of actuators required for a good vibration reduction. For the free–free suspended plate set-up, one actuator in a corner of the plate is sufficient, because all modes can be excited in this position. Also in the considered practical applications, one actuator is often sufficient to suppress the most efficiently radiating resonance frequencies, e.g. in a punching machine one actuator in a corner of the frame can control the most efficiently radiating structural modes.
The second part in the definition of the control configuration is the selection of the sensor locations. For arbitrary sensor positions, resonances as well as anti-resonances appear in the transfer function between the actuator and the sensor. The resonance frequencies are identical for all the possible sensor locations, while the anti-resonance frequencies strongly depend on the sensor position. At a resonance frequency, the plate response to a force is dominated by one mode shape and a large vibration reduction at the error sensor will result in a global reduction of the plate vibrations. At an off-resonance frequency, the plate response is governed

by several modes, and a good vibration reduction at the error sensor will not necessarily result in a global reduction. Especially at an anti-resonance frequency of the actuator–sensor transfer function, a high force will be generated by the secondary actuator to create a good vibration reduction at the sensor location. However, since an anti-resonance is not a global property of the plate, this high force will cause high vibration levels at the greater part of the plate and the global vibration level will be amplified instead of being reduced.

It is clear that, when there are anti-resonances in the actuator–sensor transfer function, the local impact disturbance rejection at the error sensor does not result in a good global performance. This problem can be solved by an optimal choice of the control algorithm or by a suitable selection of the control configuration. A control algorithm should be developed which only cancels the error sensor vibrations at the resonance frequencies without sending a control signal at the intermediate anti-resonance frequencies. In this way, the resonance frequencies can be damped, resulting in a global vibration reduction at these frequencies, while the response at the anti-resonance frequencies is unaltered. Unless a very specific collocated control configuration is used (see the next section), the required control algorithm will be very complicated for systems with a high modal density in the controlled frequency range. A second solution is the avoidance of anti-resonances in the transfer function between the actuator and the sensor. The position of the error sensor on the plate is defined according to this second solution. Since most of the acoustic energy is radiated between 500 and 900 Hz (Section 2), the ASAC strategy should focus on a noise reduction in this frequency range. Therefore the error sensor should be positioned such that, between 500 and 900 Hz, all the efficiently radiating modes can be observed and the number of zeros in the transfer function between the actuator force and the sensor response is as small as possible. Forty possible error sensor positions have been compared experimentally. After measuring the transfer functions between the secondary actuator and all these possible sensor positions, the observability of the efficiently radiating modes between 500 and 900 Hz is checked for each sensor. The sensor locations with a limited observability of one or more modes are no longer retained. Afterwards, the number of zeros between 500 and 900 Hz are defined for the remaining transfer functions. A few locations can be found with only one zero in this frequency range. Out of these locations, the sensor, corresponding with the highest minimum value of the transfer function between 500 and 900 Hz, is selected. This position, which is used in the practical experiments, is also indicated in Fig. 1. The transfer function between the actuator force and the sensor response at this position is shown in Fig. 3(a). A more detailed representation of the frequency range between 500 and 900 Hz is plotted in Fig. 3(b): it is clear that all the efficient radiating modes are present and that only one zero around 800 Hz (with sufficient damping) occurs in this transfer function. Consequently, if a control algorithm can be designed, which cancels the disturbance at this error sensor, this will result in a global reduction of the radiated noise. The design of such algorithm is the subject of the following section.

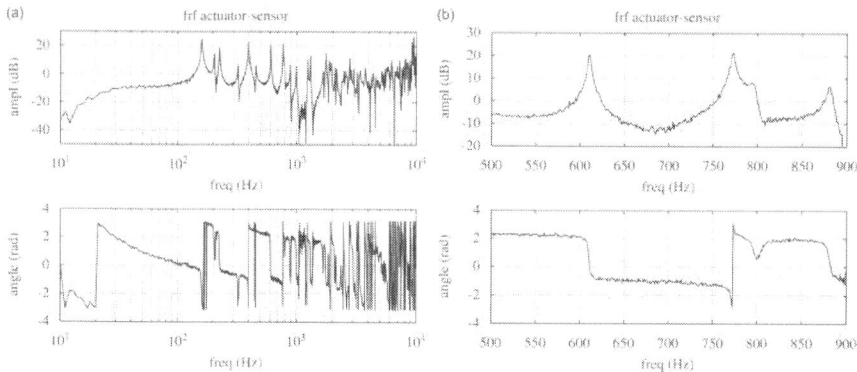

Figure 3. The transfer function between the actuator force and the sensor: (a) over the whole frequency range and (b) in the frequency range 500–900 Hz, where the greatest noise reduction is required.

ITERATIVE LEARNING CONTROL

Introduction

The second step in the design of an ASAC system, after the selection of a suitable control configuration, is the implementation of an efficient controller, which processes the sensor information to send a control signal to the actuator(s). Because the conventional feedback and feedforward ASAC control algorithms for the reduction of stationary noise are not suitable in this application (due to respectively the high modal density of the radiating structure and the transient character of the disturbance), an iterative learning control (ILC) algorithm is applied in the presented ASAC controller for the plate. ILC is a relatively new area of study in control theory, very suitable to cancel repetitive disturbances [16].

The ILC approach was motivated by the intuitive idea that it should be possible to improve the performance of a system that performs repetitively the same task and reproduces continuously the same error (welding robots, pick-and-place machines, etc.). Using the experience from the past, modifications to the input signal can be applied to the system during the next operation in order to obtain a better future performance. The first contribution on ILC, a paper by Uchiyama [17], was not widely known, because it was only published in Japanese. The idea of ILC was further developed by Arimoto [18] and [19], mainly active in the field of robotics, and became also popular among other researchers (Longman [20], Horowitz [21], etc.). A good survey of ILC can be found in the book of Moore [22] and a more recent overview paper [23]. Longman [20] presents some practical ILC design rules for engineering applications. In a recent paper by Goldsmith [24] it was shown that ILC was not fundamentally different from time-invariant control methods. He proves that causal ILC can do no better than conventional feedback control and suggests that future work on ILC should focus on the benefits of noncausal ILC filters. The practical usefulness of a noncausal ILC system will be demonstrated extensively in this section. Although the potential of ILC was demonstrated for a broad class of applications during the last two decades, the technique has only very recently been used in the field of ANC [25] and ASAC [26]. In both papers, the results obtained by the ILC controller are superior compared to those achieved by conventional feedback and feedforward control methods.

Theoretical Background

The plate system, that is controlled in this paper, is considered to be linear time-invariant and causal and is described by the following system description:

$$y(k) = P_{sec}(D)u(k) + P_{dist}(D)d(k) \qquad (1)$$

With

D	the unit delay operator ($Dx(k)=x(k+1)$)
$P_{sec}(D)$	the discrete operator, which defines the relation between the input signal to the secondary actuator and the signal measured by the error accelerometer
$P_{dist}(D)$	the discrete operator, which defines the relation between disturbing impact force and the signal measured by the error accelerometer
y(k)	the signal measured by the error accelerometer at time interval k
u(k)	the control signal sent to the secondary actuator at time interval k
d(k)	the disturbing impact force at time interval k

Since separate impacts are studied, which generate transient vibrations in the plate, several discrete time intervals with a fixed duration (p time steps) can be studied separately. A trigger signal, which announces a new impact and defines the beginning k_i of a new time interval, is supposed to be available. Define the p step histories of the error signal, the control signal and the disturbance force at the ith impact according to:

$$\mathbf{y}_i = [y_i(0) \quad y_i(1) \quad \ldots \quad y_i(p-1)]^T \tag{2}$$

$$\mathbf{u}_i = [u_i(0) \quad u_i(1) \quad \ldots \quad u_i(p-1)]^T \tag{3}$$

$$\mathbf{d}_i = [d_i(0) \quad d_i(1) \quad \ldots \quad d_i(p-1)]^T \tag{4}$$

with

$$y_i(k) = y(k_i + k) \tag{5}$$

$$u_i(k) = u(k_i + k) \tag{6}$$

$$d_i(k) = d(k_i + k) \tag{7}$$

Because the successive disturbing impact forces are supposed to be repetitive, all the time series d_i are equal to $\mathbf{d}$. Using Eqs. (2), (3) and (4), the system description Eq. (1) can be posed in a matrix form:

$$y_i = P_{sec} u_i + P_{dist} d_i \tag{8}$$

with

$$P_{sec} = \begin{bmatrix} h_{P_{sec}}(0) & 0 & \cdots & 0 \\ h_{P_{sec}}(1) & h_{P_{sec}}(0) & & \vdots \\ \vdots & & \ddots & 0 \\ h_{P_{sec}}(p-1) & h_{P_{sec}}(p-2) & \cdots & h_{P_{sec}}(0) \end{bmatrix} \tag{9}$$

and

$$\mathbf{P}_{dist} = \begin{bmatrix} h_{P_{dist}}(0) & 0 & \cdots & 0 \\ h_{P_{dist}}(1) & h_{P_{dist}}(0) & & \vdots \\ \vdots & & \ddots & 0 \\ h_{P_{dist}}(p-1) & h_{P_{dist}}(p-2) & \cdots & h_{P_{dist}}(0) \end{bmatrix} \tag{10}$$

where $h_{Psec}(k)$ and $h_{Pdist}(k)$ are the discrete impulse responses of the operators $P_{sec}(D)$ and $P_{dist}(D)$.

In this paper, a first order, trial-invariant ILC algorithm, which is represented in Fig. 4, is applied. In first-order ILC, the control signal for a new impact u_i only depends on the control signal u_{i-1} and the remaining error signal y_{i-1} at the previous impact, while in higher-order ILC [27] and [28] the control and error signals of earlier impacts u_{i-2}, y_{i-2}, u_{i-3}, y_{i-3}, etc. can also influence the control signal u_i. The control filters are supposed to be trial-invariant, which means that the control law between the input signals u_{i-1} and y_{i-1} and the calculated ILC control signal u_i is invariant over all impacts. The ILC control law, applied in the developed control algorithms, is linear: the control signal for the new impact u_i is a linear combination of the previous control signal u_{i-1} and the error signal y_i, measured at the previous impact:

$$u_i = Q u_{i-1} + L y_{i-1} \tag{11}$$

with

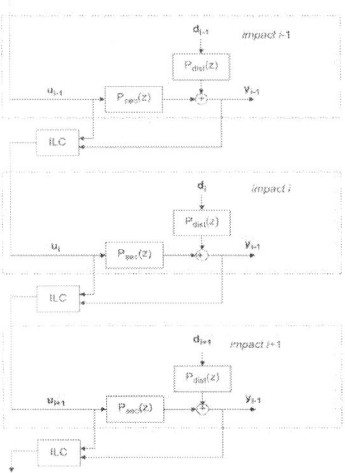

Figure 4. Control scheme of first order, trial-invariant ILC.

Q, L: p ×p-matrices

In most applications, $\mathbf{Q}$, which relates the control signal for the new impact u_i to the previous control signal u_{i-1}, is a diagonal matrix with constant coefficients, mostly equal to 1. $\mathbf{L}$, which defines the relation between the new control signal u_i and the error signal at the previous impact y_{i-1}, determines the stability of the ILC algorithm. In case of causal ILC, $\mathbf{L}$ is lower triangular, which means that a time sample of the new control signal $u_i(n)$ is only influenced by earlier samples of the error at the previous impact $y_{i-1}(n)$, $y_{i-1}(n-1)$, $y_{i-1}(n-2)$, etc. Since ILC is an offline method, which processes the data from a previous impact to calculate the control signal for the new impact, also noncausal operators can be used in the design of ILC filters. Fig. 5 shows how these noncausal operators can use 'future' time samples from a previous impact $(y_{i-1}(n+1)$, $y_{i-1}(n+2)$, etc.) to calculate the current control action. Due to this noncausal filtering, the ILC controller can anticipate for a new impact, announced by the trigger signal. In this paper, the benefits of noncausal ILC are investigated, allowing $\mathbf{L}$ to be a full matrix of learning gains.

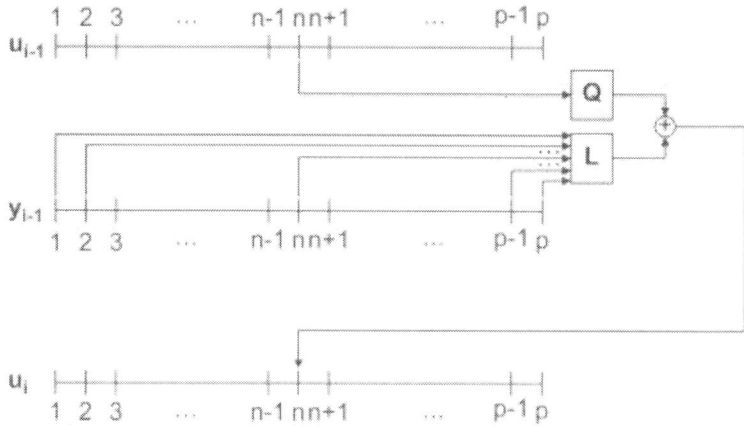

Figure 5. The use of a noncausal operator L in the ILC design.

The ILC system, described by Eq. (11), will be bounded-input, bounded-output stable, if the magnitudes of all the eigenvalues of $\mathbf{Q}$-$_{Psec}\mathbf{L}$ are less than 1 [29]:

$$|\lambda_s(\mathbf{Q} - \mathbf{P}_{sec}\mathbf{L})| < 1 \quad \forall s \qquad (12)$$

When this condition is fulfilled, the error signal converges to

$$y\infty=(\mathbf{I}-\mathbf{Q}-_{\mathrm{Psec}}\mathbf{L})^{-1}(\mathbf{I}-\mathbf{Q})_{\mathrm{Pdist}}\mathbf{d} \tag{13}$$

It is important to notice that the error signal can only become 0 if $\mathbf{Q}=\mathbf{I}$. This is the reason why most of the proposed ILC schemes operate with $\mathbf{Q}=\mathbf{I}$.

In the remainder of this paper, $\mathbf{Q}$ is supposed to be diagonal with constant coefficients q and $\mathbf{L}$ is supposed to be the matrix representation of a noncausal time-invariant difference equation with transfer function $L(z)$:

$$\mathbf{Q}=q\mathbf{I} \tag{14}$$

and

$$\mathbf{L}=\begin{bmatrix} h_L(0) & h_L(-1) & h_L(-2) & \cdots & h_L(-(p-1)) \\ h_L(1) & h_L(0) & h_L(-2) & & h_L(-(p-2)) \\ h_L(2) & h_L(1) & h_L(0) & & h_L(-(p-3)) \\ \vdots & & & \ddots & \vdots \\ h_L(p-1) & h_L(p-2) & h_L(p-3) & \cdots & h_L(0) \end{bmatrix} \tag{15}$$

where $h_L(k)$ is the discrete impulse response of $L(z)$.

Analogous to this expression, $\mathbf{Q}$ can be interpreted as the matrix representation of a difference equation with transfer function $Q(z)=q$.

Although stability and convergence can be obtained by fulfilling condition Eq. (12), the practical usefulness of this formula is very limited. Since a monotonic convergence is not guaranteed, the transient behaviour of the ILC algorithm can be undesirable. The poor transient behaviour of some ILC algorithms, although complying to Eq. (12), is clearly illustrated in Ref. [20]: it is possible that a disturbance has first grown to undesirably high levels during the first impacts before it is ultimately canceled after a high number of impacts. To assure good transients, a criterion for monotonic convergence is required. The following condition, formulated in the frequency domain, guarantees a monotonic decay of the amplitudes of all the frequency components [20]:

$$|q-L(e^{i\omega T})_{\mathrm{Psec}}(e^{i\omega T})|<1 \tag{16}$$

With

T	the sampling time
ω	the angular frequency
$L(e^{i\omega T})$	the discrete Fourier transform of L
$P_{sec}(e^{i\omega T})$	the discrete Fourier transform of P_{sec}

This criterion indicates that the Nyquist curve of $L(e^{i\omega T})P_{sec}(e^{i\omega T})$ has to be located inside a unit circle with the center point $(q,0)$, commonly denoted as the learning circle. Although this formula is only correct for a steady-state response, Longman shows in Ref. [20] that it is also a good condition to get a reduction of the transient response.

In the design of the ILC control matrices $\mathbf{Q}$ and $\mathbf{L}$, developed for the control of the plate, a trade-off has to be made between performance (Eq. (13)) and stability (Eq. (16)). Therefore the following design procedure is suggested:

1. The transfer function between the secondary actuator and the error sensor is measured.
2. The control filter $L(z)$ is shaped such that, in the frequency range where control performance is required,
   - The phase of $L(e^{i\omega t})P_{sec}(e^{i\omega t})$ stays between $-90°$ and $90°$ to assure stability (Eq. (16)with $q \approx 1$) and
   - The maximum amplitude of $L(e^{i\omega t})P_{sec}(e^{i\omega t})$ is almost equal to 1 to guarantee performance (Eq. (13)).

3. A bandpass filter is added to the $L(z)$-filter for stability reasons such that the amplitude of$L(e^{i\omega T})P_{sec}(e^{i\omega T})$ decreases significantly outside the frequency range of interest, where the$-90°$ and $90°$-phase limits are exceeded.
4. The last phase in the ILC design, is the definition of q, which should be chosen as close as possible to 1 for a good performance (Eq. (13)): if for $q=1$ the Nyquist curve of $L(e^{i\omega T})P_{sec}(e^{i\omega T})$ goes outside the learning circle, a slightly smaller value should be chosen for q to introduce robustness in the control algorithm.

Practical Implementation
Two ILC algorithms are discussed in the remainder of this section. First a causal ILC algorithm is presented to show the equivalence of this algorithm with time-invariant feedback control algorithms and to indicate

the limitations in the design of the ILC controller when only causal filters are used. In the second example, the interesting benefits of noncausal ILC filters are exploited. In all the experiments, the secondary actuator is an inertia shaker and the error sensor is an accelerometer.

Example 1: Causal ILC

Due to the high modal density of the plate in the controlled frequency range, it is impossible to design a stable causal control algorithm of limited order for the optimal control arrangement, defined in the previous section. Therefore, the causal algorithm is implemented for a collocated control configuration, where the actuator as well as the sensor are placed in the same corner of the plate. This collocated configuration with a dual actuator and sensor pair results in some very attractive stability properties, i.e. alternating poles and zeros in the secondary plant transfer function P_{sec} (Fig. 6).

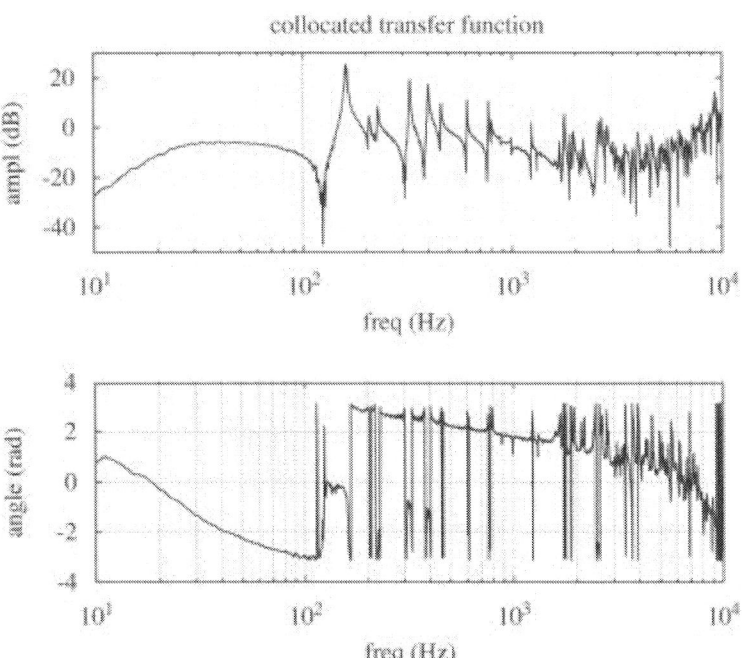

Figure 6. The transfer function P_{sec} between the collocated actuator force and the accelerometer signal.

The control matrices $\mathbf{Q}$ and $\mathbf{L}$ are designed according to the aforementioned methodology. First, the L-filter was shaped. In the lower frequency range, the phase of P_{sec} can be easily compensated by a negative phase of the L-filter to keep the phase of LP_{sec} between $-90°$ and $90°$. The gain of L is adjusted such that the maximum amplitude of LP_{sec} is close to 1 in this lower-frequency range. In the higher frequency range (above 500 Hz), it is impossible to design a causal L-filter, which compensates the phase loss of P_{sec} caused by a time delay. Consequently, for robustness, a lowpass filter should be introduced in L to decrease the amplitude of LP_{sec} at the higher frequencies, where the phase of LP_{sec} exceeds the $(-90°, 90°)$-limits. This results in the following L-filter (Fig. 7(a)):

$$L = \frac{-30}{s + 2\pi 60} \tag{17}$$

This filter was developed in the continuous domain and afterwards transformed to the discrete domain using Tustin's transformation rule. Since for $q=1$ the Nyquist curve of LP_{sec} exceeds the learning circle, q has to be reduced to 0.97. Fig. 7(b) shows that for this value of q the stability condition for monotonic convergence Eq. (16) is fulfilled.

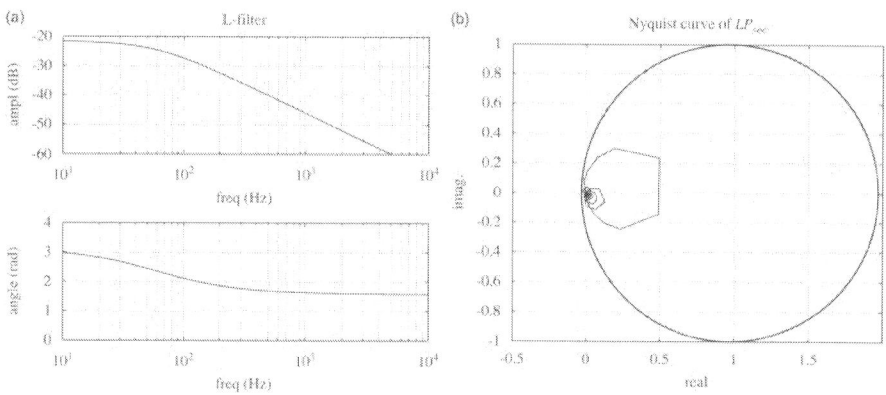

Figure 7. The design of the control filters Q and L: (a) the transfer function of the L-filter and (b) the Nyquist plot of LP_{sec} inside the learning circle (the stability test for $q=0.97$).

The reduction $|Y_\infty(e^{i\omega T})|/|P_{dist}(e^{i\omega T})D(e^{i\omega T})|$, which can be theoretically obtained by the implementation of the developed ILC algorithm, is evaluated using Eq. (13) and is shown in Fig. 8($Y_\infty(e^{i\omega T})$ and $D(e^{i\omega T})$ are the discrete Fourier transform of y_∞ and d, respectively). The greatest theoretical reduction can be achieved at the resonance frequencies of the plate, while there is no reduction at the intermediate frequencies. According to Eq. (13), the final error at the sensor will be only small, if $P_{sec}L$ is high compared to $I-Q$. Since the alternating poles and zeros of P_{sec} are not compensated by the L-filter, $P_{sec}L$ will only be high enough at the poles of P_{sec} to create a vibration reduction. At the intermediate frequencies between the poles, no control signal will be sent to the secondary actuator and the vibration pattern is unaltered. Due to the lowpass characteristic of the ILC filter L, the frequency band in which reduction can be obtained at the error sensor is limited: the error can only be reduced by more than 5 dB at the resonance peaks of the plate below 450 Hz. It is also important to notice that this low-frequency performance does not result in a deterioration in the higher frequency range.

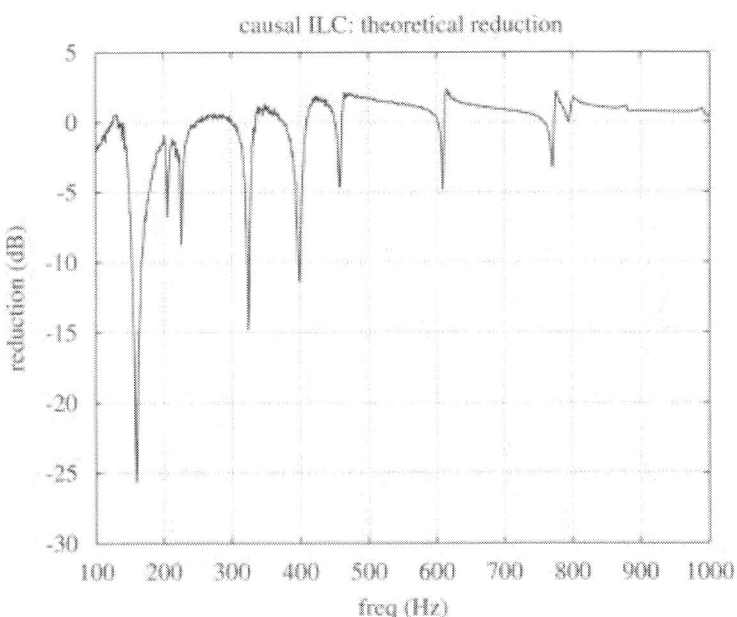

Figure 8. The theoretical reduction $|Y_\infty(e^{i\omega T})|/|P_{dist}(e^{i\omega T})D(e^{i\omega T})|$, which can be achieved by the causal ILC controller as well as by the equivalent time-invariant feedback controller.

Goldsmith [24] and Verwoerd [30] state that for causal ILC filters Q and L an equivalent time-invariant feedback controller exists, which yields the same theoretical reduction at the error sensor (Fig. 8). This equivalent feedback controller can be defined as

$$C_{FB} = (I - Q)^{-1} L \tag{18}$$

In this application, the resulting equivalent controller becomes

$$C_{FB} = \frac{L}{1 - q} = \frac{-30}{0.03(s + 2\pi 60)} \tag{19}$$

The performance of the ILC controller is practically compared with the time-invariant feedback controller on the plate set-up. Both algorithms are implemented on a dSPACE 1103 DSP board, which calculates the control signals for the secondary actuator. Fig. 9(a) shows the error signals in the time domain, when the plate is excited by a primary impact force. The different curves compare the remaining error signals, which are obtained by the ILC algorithm after a different number of impacts. During the first 80 impacts, the error acceleration, caused by the primary shaker, becomes significantly smaller at each impact, due to the updated ILC control force. This learning behaviour can also be observed in Fig. 9(b): there is a clear reduction of the error signal in the lower frequency range between 50 and 250 Hz during the first 80 impacts. After the 80th impact, the ILC algorithm has converged and there is no further improvement at the consecutive pulses. While the vibration reduction is significant in the lower frequency, the ILC algorithm has no influence on the error signal in the higher frequency range. This behaviour was exactly predicted by the theoretical formula in Eq. (13), as was explained above.

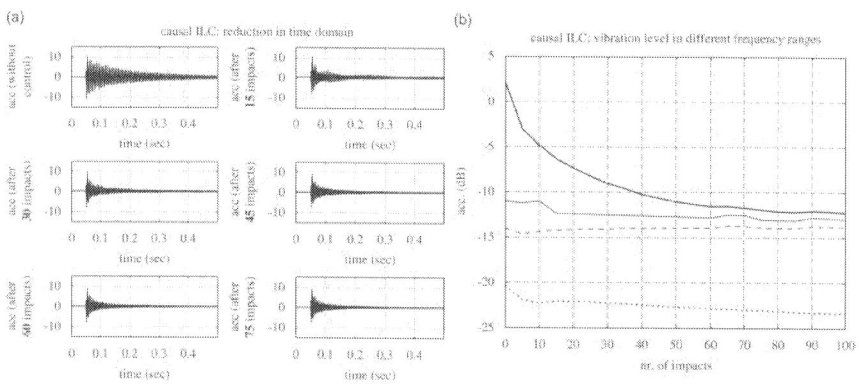

Figure 9. The performance of the causal ILC controller: (a) the error acceleration after a certain number of learning impacts (shown in the time domain) and (b) the error acceleration as a function of the number of learning impacts (shown in different frequency bands: 50–250 Hz (bold line), 250–500 Hz (dotted line), 500–750 Hz (solid line) and 750–1000 Hz (dashed line)).

Contrary to the ILC controller, the performance of the time-invariant feedback controller is equal for all the controlled impact disturbances. The reduction achieved by this time-invariant controller is plotted in Fig. 10 and compared to the reduction obtained by the ILC controller after convergence. It is clear that the time-invariant feedback controller can achieve the same error after one impact as the ILC controller obtains only after convergence, which requires a large number of learning impacts. This result shows that it is useless to apply a causal learning controller, which requires a certain number of impacts to update the control signal before convergence is achieved. Another disadvantage of the ILC algorithm, which can be avoided by applying a time-invariant feedback algorithm, is the necessity of an accurate trigger signal to predict the next impact: while the time-invariant controller is permanently active, the ILC controller only sends a control signal to the secondary actuators, when a new impact is announced by a trigger signal. Since this example shows that there is no reason to use causal ILC instead of time-invariant feedback control, in the following paragraphs the possibilities of noncausal ILC are investigated.

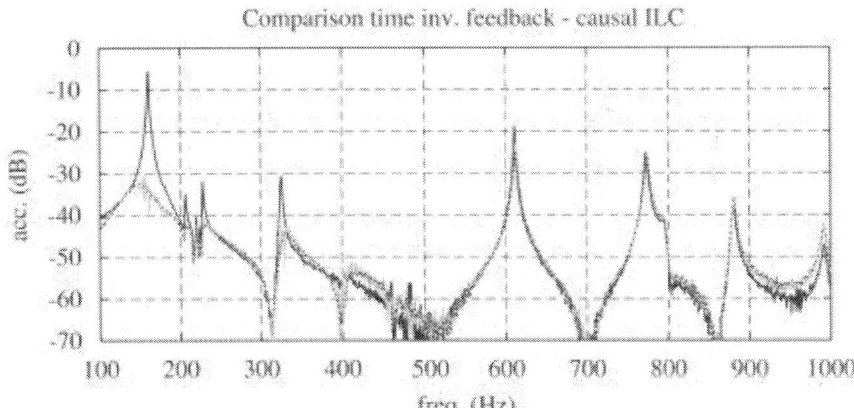

Figure 10. The remaining error signal, measured by the error accelerometer, without control (bold line), with feedback control (dotted line) and with ILC after convergence (grey line).

Example 2: Noncausal ILC

In this paragraph, a noncausal ILC algorithm is designed for the control of the optimal sensor–actuator configuration, which was selected in Section 3. The transfer function of the secondary plant P_{sec} for this control arrangement is shown in Fig. 3(a). The control matrices Q and L are again designed according to the methodology, presented in the first part of this section. The frequency range, where control performance is desired, is situated between 500 and 900 Hz. Since in this range the phase of LP_{sec} should stay between $-90°$ and $90°$, the phase lag of $540°$ in P_{sec} between 500 and 900 Hz caused by 4 poles and only 1 zero should be compensated by 3 zeros in the control filter L, which are placed at 620, 780 and 880 Hz. In order to limit the high-frequency amplification of L, each zero in the controller is compensated by a pole. Therefore 3 poles at frequencies higher than 900 Hz are introduced in L (1 pole at 900 Hz and 2 poles at 5000 Hz). This results in the following controller L_1:

$$L1 = L_{1,a} L_{1,b} L_{1,c} \tag{20}$$

with

$$L_{1,a} = \frac{900^2\, s^2 + 2 \times 0.01(2\pi620)s + (2\pi620)^2}{620^2\, s^2 + 2 \times 0.03(2\pi900)s + (2\pi900)^2} \tag{21}$$

$$L_{1,b} = \frac{5000^2\, s^2 + 2 \times 0.01(2\pi780)s + (2\pi780)^2}{780^2\, s^2 + 2 \times 0.03(2\pi5000)s + (2\pi5000)^2} \tag{22}$$

$$L_{1,c} = \frac{5000^2\, s^2 + 2 \times 0.03(2\pi880)s + (2\pi880)^2}{880^2\, s^2 + 2 \times 0.03(2\pi5000)s + (2\pi5000)^2} \tag{23}$$

Due to the 3 poles between 500 and 900 Hz, the phase of LP_{sec} stays inside a band of $180°$ in this frequency range. To shift this $180°$-band between $-90°$ and $90°$, a second compensator has been added to L_1:

$$L_2 = L_1 \frac{2\pi50 \quad s^2 + 2 \times 0.4(2\pi1000)s + (2\pi1000)^2}{s + 2\pi50\, s^2 + 2 \times 0.4(2\pi2000)s + (2\pi2000)^2} \tag{24}$$

Figs. 11(a and b) show the transfer functions of L_2 and L_2P_{sec}. It is clear that, in the frequency range of interest (500–900 Hz), the phase of L_2P_{sec} stays between $-90°$ and $90°$. The developed L_2-filter is still causal, the benefits of noncausality are only used in the second phase of the controller design. Two noncausalbandpass filters L_3 and L_4 are added to L_2 to create L such that the amplitude of LP_{sec} decreases significantly outside the frequency range of interest without any phase change. L_3 consists of a noncausallowpass filter $L_{3,lp}$ and a noncausalhighpass filter $L3_{,hp}$. Since a strong roll-off is necessary to decrease the amplitude of LP_{sec} sufficiently outside the frequency range 500–900 Hz, these both filters are of 12th order (the actual implementation of these noncausal filters is explained inAppendix A). In both filters, the phase lag due to a causal pole is compensated by the phase lead due to an anti-causal pole (an anti-causal system does not process earlier samples from a previous impact but only uses future time samples to calculate the current control action):

$$L3 = L3_{,lp} L3_{,hp} \tag{25}$$

with

$$L_{3,lp} = \left(\frac{(2\pi600)^4}{(s^2 + 2 \times 0.4(2\pi600)s + (2\pi600)^2)(s^2 - 2 \times 0.4(2\pi600)s + (2\pi600)^2)} \right)^3 \tag{26}$$

$$L_{3,hp} = \left(\frac{s^4}{(s^2 + 2 \times 0.4(2\pi800)s + (2\pi800)^2)(s^2 - 2 \times 0.4(2\pi800)s + (2\pi800)^2)} \right)^3 \tag{27}$$

Finally, a second noncausal compensator $L4$ was also implemented in the ILC control filter L ($=L2L3L4$) to create some extra robustness. The compensator $L4$ consists of 2 causal (both at 700 Hz) and 2 anti-causal (also at 700 Hz) poles as well as 2 causal (at 400 and 1200 Hz) and 2 anti-causal zeros (also at 400 and 1200 Hz). This further increases the amplitude of LP_{sec} between 500 and 900 Hz and reduces its amplitude just below and above this frequency range, again without any phase change.

$$L_4 = \frac{L_{zero,1} L_{zero,2}}{L_{pole}^2} \tag{28}$$

with

$$L_{zero,1} = (s^2 + 2 \times 0.05(2\pi400)s + (2\pi400)^2)(s^2 - 2 \times 0.05(2\pi400)s + (2\pi400)^2) \tag{29}$$

$$L_{zero,2} = (s^2 + 2 \times 0.1(2\pi1200)s + (2\pi1200)^2)(s^2 - 2 \times 0.1(2\pi1200)s + (2\pi1200)^2) \tag{30}$$

$$L_{pole} = \frac{1}{(s^2 + 2 \times 0.3(2\pi 700)s + (2\pi 700)^2)(s^2 - 2 \times 0.3(2\pi 700)s + (2\pi 700)^2)}$$

(31)

The gain of L is adjusted such that the maximum amplitude of LP_{sec} is close to 1 in the frequency range of interest. The transfer function of the resulting L-filter is shown in Fig. 12(a). Compared to the transfer function of $L2$, there is no change of the phase due to the use of noncausal filters, such that the phase of LP_{sec} still stays between $-90°$ and $90°$ between 500 and 900 Hz. However, the amplitude is strongly reduced below 500 Hz and above 900 Hz, generating the desired bandpass characteristic. If the same bandpass filtering of the amplitude would have been created by a causal filter, this would have caused a strong phase change over a broad frequency range, also between 500 and 900 Hz, the frequency range of interest. It is impossible to keep the phase of LP_{sec} between $-90°$ and $90°$ in this latter range with a causal ILC controller, that generates the required bandpass characteristic of L. This bandpass filtering without a phase change is a great benefit of noncausal filters.

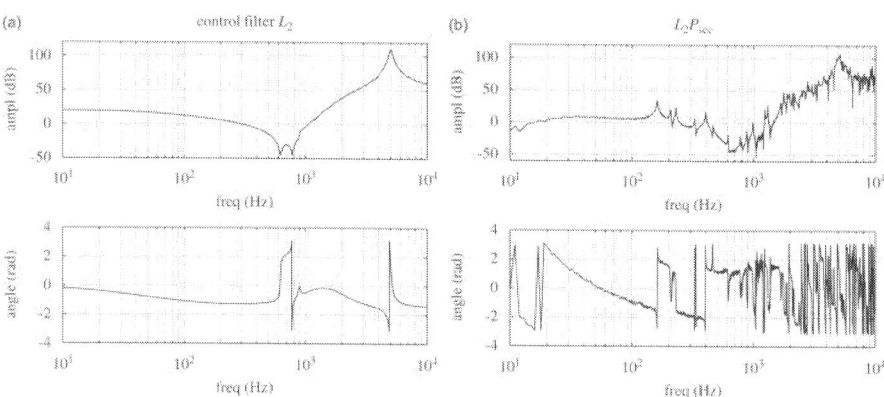

Figure 11. The first part in the design of the ILC filter for the control of the optimal control configuration: the transfer functions of (a) the ILC control filter $L2$ and (b) $PsecL2$.

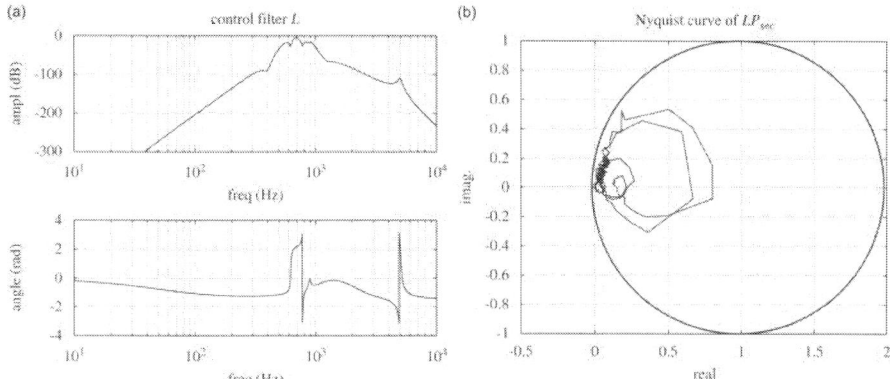

Figure 12. The second part in the design of the ILC filter for the control of the optimal control configuration: (a) the transfer function of the total ILC control filter L and (b) the Nyquist plot of LP_{sec} inside the learning circle (the stability test for $q=0.98$).

Based on the selected L-filter, the value q, which defines the Q-filter, is chosen. Since for a q-value of 1 the condition for monotonic convergence Eq. (16) is not fulfilled, q is reduced to 0.98. Fig. 12(b) shows that for this value of q the Nyquist curve of LP_{sec} does not exceed the learning circle, which guarantees monotonic convergence.

The theoretical reduction $|Y_\infty(e^{i\omega T})|/|P_{dist}(e^{i\omega T})D(e^{i\omega T})|$, which can be achieved by the selected L- and Q-filter, is shown in Fig. 13. Contrary to the causal controller in the previous section, a high error reduction can be achieved by this noncausal ILC controller in the higher frequency range, where most of the noise is radiated. The error in the error accelerometer can be reduced by at least 10 dB over the whole frequency range of interest from 500 to 900 Hz. While the causal controller only reduces the error at the resonance peaks, the noncausal controller with a L-filter, which compensates all the zeros and poles of P_{sec}, creates a broadband vibration reduction at the error sensor: there is not only a reduction at the resonance frequencies but also at the intermediate frequencies.

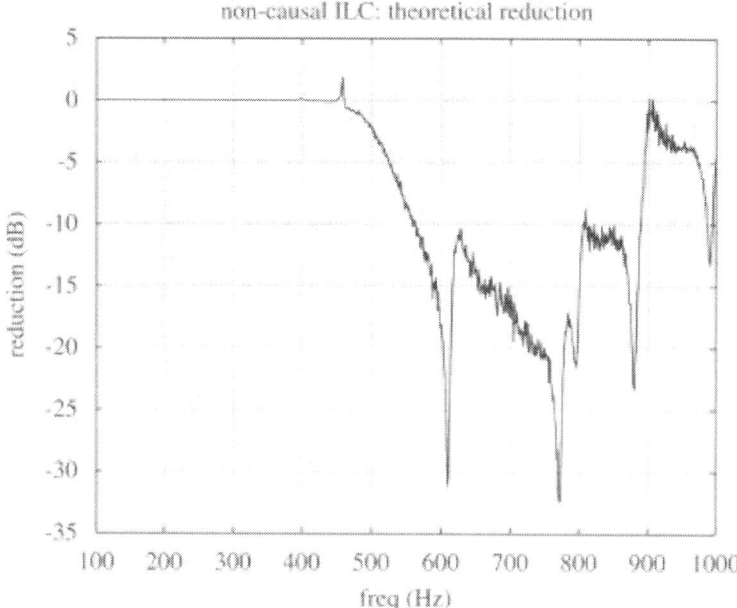

Figure 13. The theoretical reduction $|_{Y\infty}(e^{i\omega T})|/|_{Pdist}(e^{i\omega T})D(e^{i\omega T})|$, which can be achieved by the noncausal ILC controller.

The noncausal controller as discussed in the previous paragraphs has been implemented on a dSPACE 1103 DSP board and applied to the plate case study. A detailed survey of the practical implementation of a noncausal control filter can be found in Appendix A. In the remainder of this section, the local results, achieved by the noncausal controller at the error sensor, are studied. The influence of the controller on the global plate vibration level and the radiated acoustic noise level will be described in the next section.

The results, obtained by the noncausal controller at the error sensor in the plate experiments, are plotted in Figs. 14(a and b). Fig. 14(a) shows the error acceleration signals in the time domain, when the plate is excited by a primary impact force. The different curves compare the remaining error signals, which are obtained by the noncausal ILC algorithm after a different number of impacts. Since most of the vibration power in the error sensor is situated in the lower frequency range below 250 Hz and the greatest reduction is achieved between 500 and 900 Hz, the reduction due to the learning process can hardly be detected in the time domain. However, the learning behaviour can clearly be observed in the frequency domain (Fig. 14(b)): in the frequency bands between 500 and 750 Hz and between 750 and 1000 Hz there is a clear reduction of more than 15 dB of the error signal during the first 50 impacts, while below 500 Hz the error signal has not changed. This is exactly the performance, which was theoretically predicted (Fig. 13). After the 50th impact, the learning behaviour of the ILC algorithm has converged and there is no further reduction.

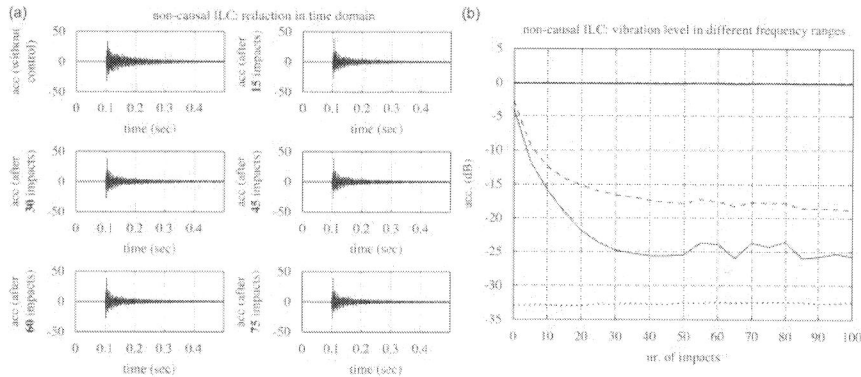

Figure 14. The performance of the noncausal ILC controller: (a) the error acceleration after a certain number of learning impacts (shown in the time domain) and (b) the error acceleration as a function of the number of learning impacts (shown in different frequency bands: 50–250 Hz (bold line), 250–500 Hz (dotted line), 500–750 Hz (solid line) and 750–1000 Hz (dashed line)).

CONTROLLER PERFORMANCE EVALUATION

This section discusses first the global performance (i.e. the total vibration level and the radiated noise level) of the implemented causal and noncausal algorithm. Afterwards the benefits of a controller, which combines both algorithms, are investigated. The global vibration level and the radiated noise level are measured, respectively, by 7 accelerometers, uniformly distributed over the plate and by 1 microphone in the far field (2 m from the plate). The global vibration level is defined as the rms-value of the accelerations measured by these 7 accelerometers.

Example 1: Causal ILC
Since the reduction in the error sensor, which is achieved by the causal ILC algorithm after convergence, is the same as the reduction by the equivalent time-invariant feedback algorithm, the global performance of

both controllers is also equal. Therefore, the results are only presented for the practically more feasible time-invariant feedback controller.

Fig. 15 shows the vibration levels of the plate without and with control. It is clear that in the lower frequency range below 500 Hz the resonance peaks are strongly damped by the causal controller. In the intermediate frequency ranges between the resonances, the global vibration level of the plate does not change, because the causal controller does not send a control signal to the secondary actuator in these frequency bands. Due to this controller design, a global vibration reduction of 10 dB is obtained between 100 and 500 Hz. In the higher frequency range above 500 Hz, the global vibration level is also not influenced by the causal controller due to the lowpassfilter in the design of L and $_{CFB}$. The resulting overall reduction of the global plate vibrations is 5.5 dB. In Fig. 16, the A-weighted radiated noise levels are plotted in the one-third octave bands below 1000 Hz. Contrary to the global vibration level, the radiated noise levels are not significantly reduced in the lower frequency range, since the low-frequency modes of a non-baffled plate are not very efficiently radiating. Especially below 250 Hz, the radiated noise level hardly exceeds the background level. However, in the frequency band of 315 Hz, where the highest amount of low-frequency $(<500\,\mathrm{Hz})$ acoustic energy is radiated, there is a clear noise reduction of 5 dB. In the higher frequency range, only a small reduction of 3 dB can be observed in the one-third octave band of 630 Hz. Due to the limited radiation efficiency of the low-frequency modes, the global plate vibration reduction results only in a 1 dB(A) total noise reduction.

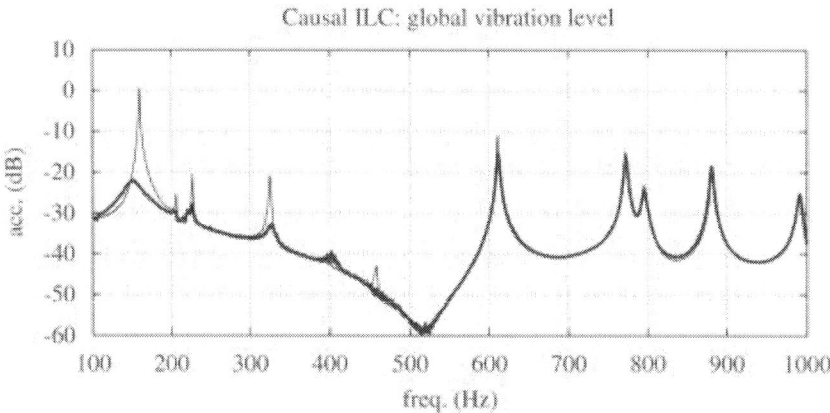

Figure 15. The global vibration level of the whole plate without control (solid line) and with causal ILC/equivalent time-invariant feedback control (bold line).

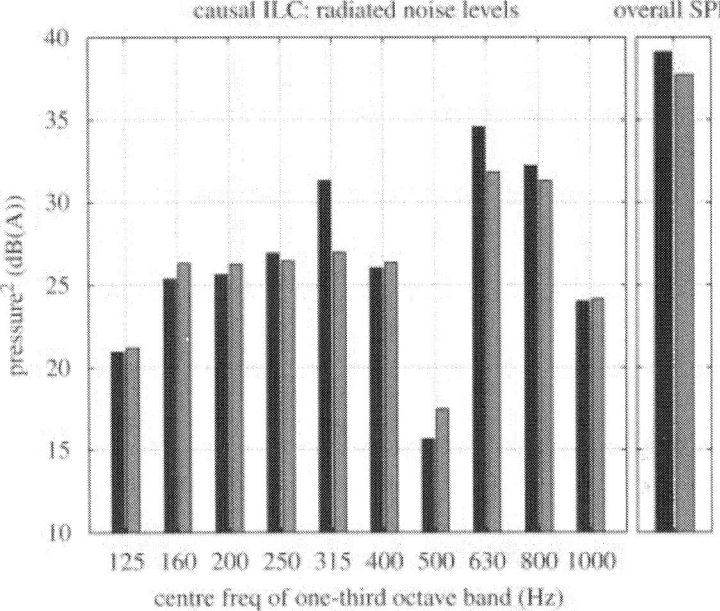

Figure 16. The A-weighted radiated noise levels in the one-third octave bands below 1000 Hz and the A-weighted overall Sound Power Level without control (black) and with causal ILC (grey).

Example 2: NoncausalILC

This paragraph studies the global effect on the plate behaviour of the noncausal controller, developed in the previous section. Contrary to the causal ILC controller, which creates a low-frequency reduction, the noncausal controller focuses on a noise and vibration reduction in the higher frequency range (500–900 Hz), where most of the radiated noise is situated. Due to the introduction of the noncausalbandpass filter in the design of L, no control signal is sent to the secondary actuator below 500 Hz. This explains why the global vibration spectrum below 500 Hz with control is the same as the uncontrolled spectrum (Fig. 17). In the previous section, Fig. 14(b) shows that between 500 and 1000 Hz the vibrations in the error sensor are significantly reduced by the control signal of the noncausal ILC controller. Due to the appropriate selection of the control configuration with successive poles in the plant transfer function $_{Psec}$ (presented inSection 3), a vibration cancellation at the error sensor is supposed to result in a global vibration reduction. This reasoning is confirmed in Fig. 17: a vibration reduction of 1.5 dB is achieved by the noncausal ILC controller in the frequency band between 500 and 1000 Hz. Because the three most efficiently radiating plate modes at 611, 772 and

881 Hz are strongly damped, the noise reduction is more pronounced than the vibration reduction: a total noise reduction of 7 dB(A) is obtained between 500 and 1000 Hz. The noise reduction in one-third octave bands is presented in Fig. 18. The overall noise reduction over the whole frequency range is only 2.5 dB(A). This level is limited by the remaining low-frequency noise especially in the one-third octave band of 315 Hz, which is not tackled by the noncausal controller.

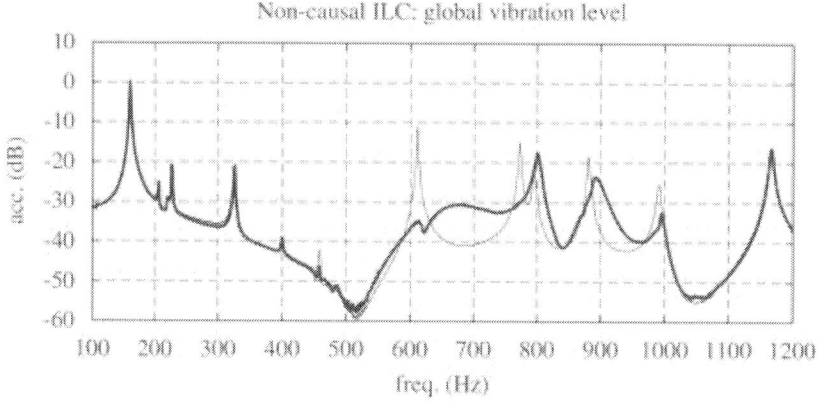

Figure 17. The global vibration level of the whole plate without control (solid line) and with noncausal ILC (bold line).

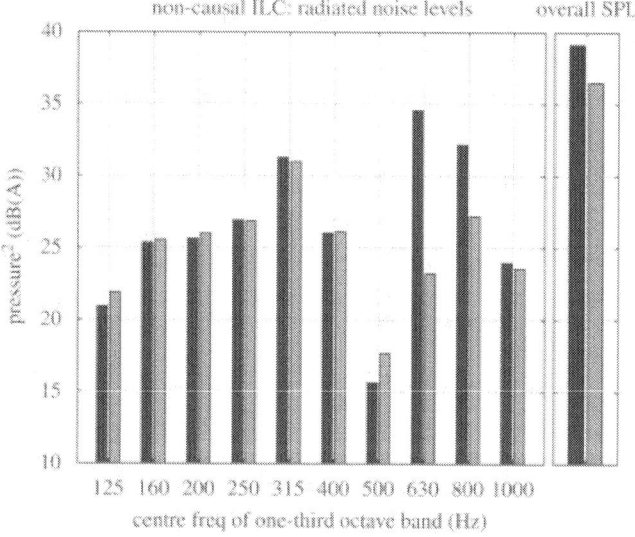

Figure 18. The A-weighted radiated noise levels in the one-third octave bands below 1000 Hz and the A-weighted overall Sound Power Level without control (black) and with noncausal ILC (grey).

Since one zero (at 800 Hz) in the transfer function between the actuator and the sensor is inevitable in the frequency range 500–900 Hz (Fig. 3(b)), there is still 1 peak at this frequency in the global vibration level with control, which limits the vibration reduction. However, this vibration peak cannot be observed in the radiated noise (Fig. 18) because the corresponding vibration mode at this frequency is not an efficient noise radiator.

It is clear that due to the good selection of the actuator and sensor position a global vibration and noise reduction can be realized in the higher frequency domain. The cancellation of the disturbance in the optimal error sensor results in a global performance. Other configurations with the error sensor at different positions, which were not optimal (i.e. no successive poles in the secondary plant transfer function P_{sec}), were also experimentally tested. Noncausal ILC control filters were designed for all these new configurations. Although a disturbance rejection could be easily obtained at the new error sensors by the controllers, in all these cases, the local disturbance reduction resulted in a higher global vibration level than the level obtained by the optimal configuration.

One of those non-optimal control configurations is briefly discussed in the following paragraph. While the actuator is located in the same corner of the plate as in the previous examples, the error sensor is placed in the middle of the plate's lower part. Figs. 19(a and b) show the transfer function of the secondary plant P_{sec} between this actuator and sensor: it is clear that two zeros with limited damping (at 620 and 778 Hz) occur in this transfer function, which is not optimal according to the procedure for the selection of the configuration developed in Section 3.

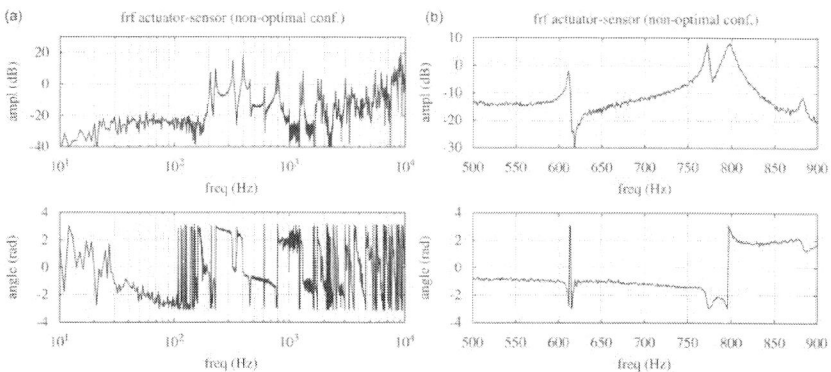

Figure 19. The transfer function between the actuator force and the sensor (non-optimal control configuration): (a) over the whole frequency range and (b) in the frequency range 500–900 Hz, where the greatest noise reduction is required.

In a similar way as in Section 4.3.2 a noncausal controller is developed for this non-optimal control configuration. The performance of the designed controller at the error sensor is plotted in Fig. 20: after convergence of the ILC algorithm a strong reduction is obtained in the frequency range of interest between 500 and 900 Hz. Although the vibration level at the error sensor is significantly reduced in this frequency band, the global vibration level (Fig. 21(a)) and the radiated noise level (Fig. 21(b)) have hardly decreased. Due to the zeros in the plant transfer function P_{sec} at 620 and 778 Hz, a high control force is necessary to cancel the vibrations at the error sensor at these frequencies. However, since a zero is not a global property of the plate, this high control force creates high vibration levels at the greater part of the plate such that the global vibration level and the radiated noise level are significantly amplified around 620 and 778 Hz. The reduction achieved by the controller at other frequencies is almost completely cancelled by the amplification at those 2 frequencies, which explains why no significant global performance can be obtained.

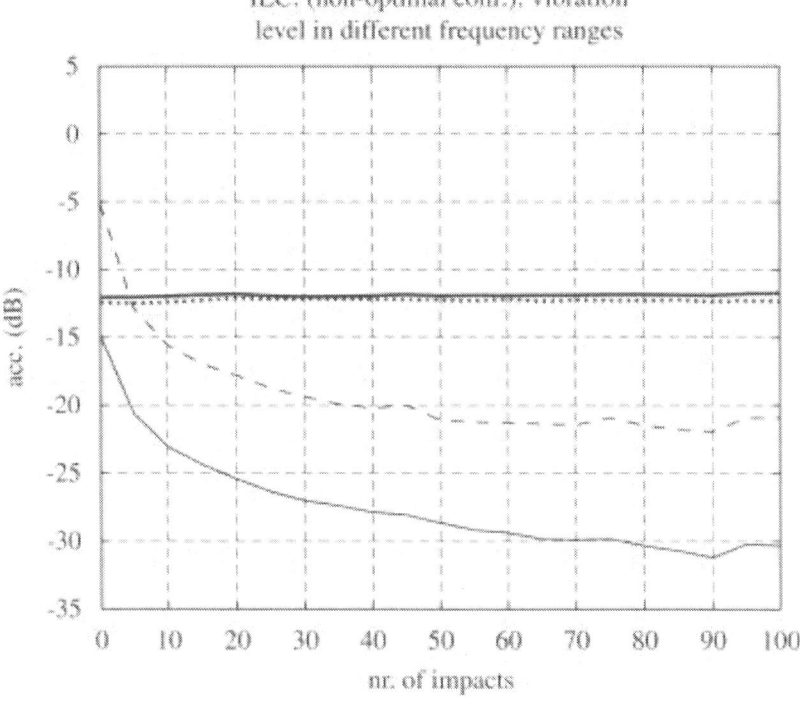

Figure 20. The performance at the error sensor of the noncausal controller (for the non-optimal control configuration): the error acceleration as a function of the number of learning impacts (shown in different frequency bands: 50–250 Hz (bold line), 250–500 Hz (dotted line), 500–750 Hz (solid line) and 750–1000 Hz (dashed line)).

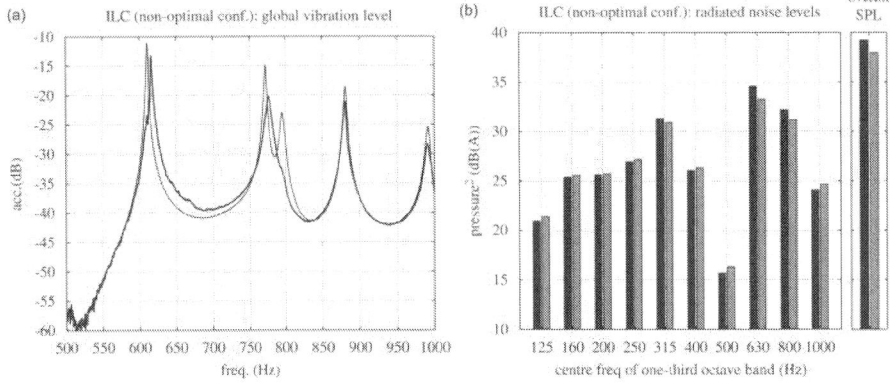

Figure 21. The global performance achieved by a controller for the non-optimal control configuration: (a) the global vibration level of the whole plate without control (solid line) and with noncausal ILC (bold line) and (b) the A-weighted radiated noise levels in the one-third octave bands below 1000 Hz and the A-weighted overall Sound Power Level without control (black) and with noncausal ILC (grey).

Example 3: Combined Time-Invariant Feedback and ILC Controller

The residual noise, which cannot be cancelled out by the noncausal controller in the second example, has a predominant low-frequency character. This low-frequency noise can be suppressed by the addition of a causal ILC controller or its equivalent time-invariant feedback controller, as presented in the first example. Therefore a control system, consisting of a time-invariant feedback controller combined with a noncausal ILC controller, is designed for the plate. The control scheme of this controller is presented in Fig. 22. The control signal u_i to the secondary actuator consists of a signal $u_{i,fb}$ from the causal feedback control filter and a signal $u_{i,ILC}$ from the ILC controller. The feedback part of the combined controller, which uses a collocated error sensor in the corner of the plate (measuring $y_{i,fb}$), uses the same control filter C_{fb} as in the first example. This controller reduces the low-frequency noise, compensates for repetitive as well as random disturbances and is immediately effective at the first controlled impact without a learning process. The noncausal ILC part of the controller is only used to reduce the residual high-frequency error, that cannot be controlled by the time-invariant feedback controller C_{fb}. The error signal for this noncausal controller $y_{i,ILC}$ is measured by a second error sensor, which is located at the optimal position to achieve global reduction: this position is identical to the optimal location defined in the second example. The development of the noncausal ILC controller (control filters Q and L) is performed in the same way as for the noncausal controller without feedback. However, the controllable plant is now $P_{ILC}/1 + C_{fb}P_{fb}$ instead of P_{ILC}, where P_{fb} and P_{ILC} are the two transfer functions between the actuator force in the corner of the plant and the responses in the error sensors, respectively.

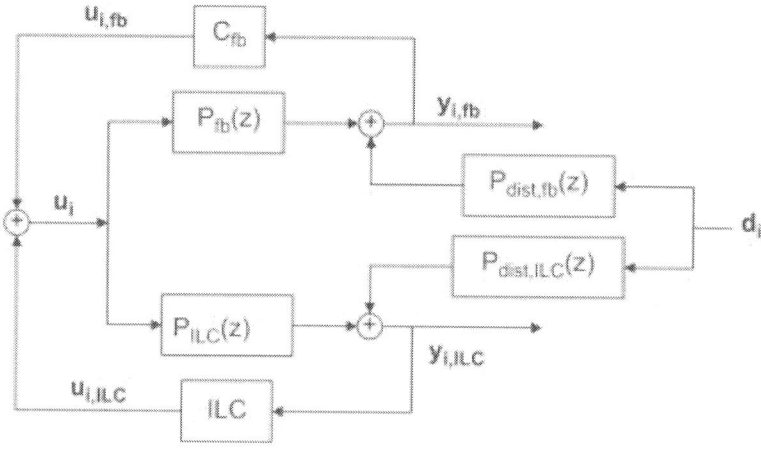

Figure 22. Control scheme of the combined feedback/ILC controller.

Fig. 23 shows the vibration levels measured by the second error sensor (measuring $y_{i,ILC}$) in 4 frequency bands below 1000 Hz during the first 100 impacts. The behaviour in the 2 frequency bands below 500 Hz differs significantly from the 2 bands above 500 Hz. Below 500 Hz, the time-invariant feedback controller is mainly active, such that no learning behaviour can be observed. The final reduction is already obtained after the first controlled impact. Above 500 Hz, the control signal from the time-invariant feedback controller strongly decreases due to the lowpass filter in the control design. In this frequency range, however, the ILC controller becomes active and will determine the control performance. In the frequency bands 500–750 Hz and 750–1000 Hz, the vibration level decreases gradually at every impact due to the learning process in the ILC algorithm. After 50 impacts, there is no further reduction, which indicates that the ILC controller has converged.

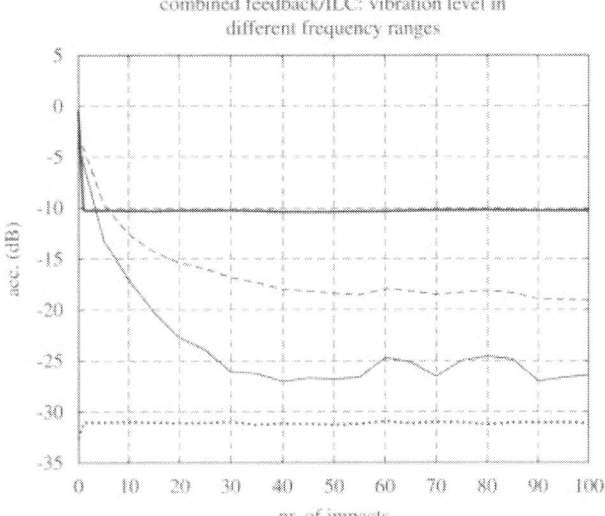

Figure 23. The performance of the combined feedback/ILC controller at the error sensor, which measures $y_{i,ILC}$: the error acceleration as a function of the number of learning impacts (shown in different frequency bands: 50–250 Hz (bold line), 250–500 Hz (dotted line), 500–750 Hz (solid line) and 750–1000 Hz (dashed line)).

The global performance of this combined controller after convergence is also measured and compared to the two other controllers. The global vibration reduction of this controller is plotted in Fig. 24. It is clear that the new controller combines the benefits of the 2 previous controllers: a vibration reduction is achieved in the lower as well as in the higher frequency range. This results in a broadband global vibration reduction of 6 dB. Fig. 25 shows the noise reduction in the one-third octave bands below 1000 Hz without and with feedback/ILC control. In the 3 one-third octave bands, where the plate radiates the highest noise levels (315, 630 and 800 Hz), significant noise reductions are obtained by this combined controller such that the total noise level is reduced by 3.5 dB(A).

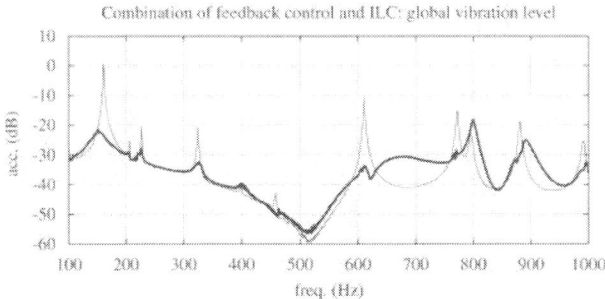

Figure 24. The global vibration level of the whole plate without control (solid line) and with combined feedback/ILC control (bold line).

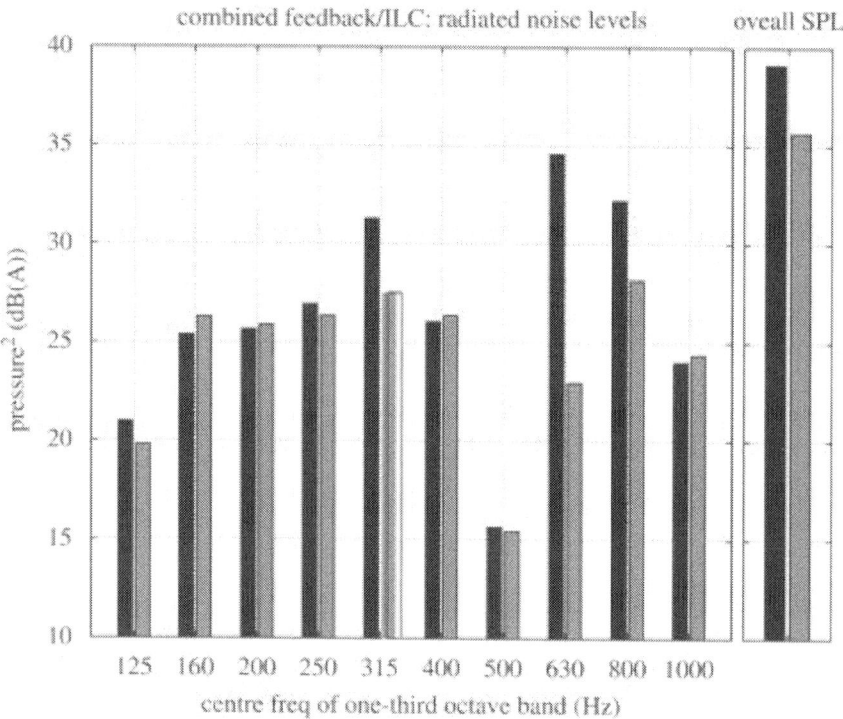

Figure 25. The A-weighted radiated noise levels in the one-third octave bands below 1000 Hz and the A-weighted overall Sound Power Level without control (black) and with the combined feedback/ILC controller (grey).

CONCLUSIONS

In this paper, a design strategy is proposed for the development of an ASAC system to reduce repetitive structural impact noise. The design process is divided into two parts: first the optimal control configuration is defined and afterwards a control algorithm is implemented for the selected actuator(s) and sensor(s). To find the optimal actuator and sensor locations, a selection procedure is developed such that the chosen control system will create a high global vibration and noise reduction in case of a perfect disturbance cancellation at the error sensors. Since an impact generates a broadband excitation, most of the noise will be radiated at certain structural resonance frequencies. Therefore in the optimal configuration for impact noise control the global vibrations are strongly reduced at these resonance frequencies without significantly amplifying the vibration levels in the intermediate frequency ranges.

In the second step of the ASAC design process, a control algorithm is developed for the selected optimal configuration. This algorithm defines the achievable reduction of the vibrations at the error sensors and consequently the possible global noise and vibration reduction. Because in many industrial applications, where the reduction of impact noise is an issue, successive impacts have a repetitive behavior, this repetitiveness can be used in the design of the control algorithms. Causal and no causal ILC controllers, which use knowledge from the previous impacts to calculate the control signal for the next impact, are presented and compared to traditional time-invariant feedback algorithms. While equivalent feedback controllers exist for causal ILC systems, this is not the case for noncausal ILC systems. Therefore, the benefits of no causal ILC algorithms (i.e. much more freedom in the design of the ILC filters) are extensively explored in this paper. The efficiency of the whole design methodology has been verified in a practical example, where an ASAC system is developed to reduce the low frequency $(< 1000 \, Hz)$ impact noise radiated by a plate. The same methodology can however also be used for the active reduction of higher frequency noise, if the higher frequency content of the successive impacts is sufficiently repetitive.

The future work consists of a theoretical and a practical part. In the first part, the possibilities of noncausal ILC algorithms will be further investigated. For example, different noncausal algorithms, each focussing on a certain frequency range, will be implemented in parallel in order to get a good performance over a broader frequency range. The second part focuses on the application of the developed methodology on real industrial machinery (e.g. punching machines, presses, etc.). In these industrial noise problems, additional challenges can be expected: the influence of the limited repetitiveness of the successive impacts has to investigated; a proper trigger signal, which announces a new impact accurately, has to be searched; probably new actuators and sensors, which can deliver and measure a high amount of energy during short periods, will be necessary; etc. However, provided these practical problems can be resolved, the good results, achieved with the presented design methodology of ASAC systems, are very promising for the use in industrial impact noise applications.

ACKNOWLEDGEMENTS

The research of G. Pinte is financed by a scholarship of the Institute for the Promotion of Innovation through Science and Technology in Flanders (IWT Vlaanderen). A part of this research is also funded by the European research project Noiseless.

REFERENCES

1. A. Akay, A review of impact noise, Journal of the Acoustical Society of America 64 (4) (1978) 977–987.
2. E.J. Richards, A. Lenzi, On the prediction of impact noise, IX: the noise from punch presses, Journal of Sound and Vibration 103 (1) (1985) 43–81.
3. W. Dehandschutter, The Reduction of Structure-borne Noise by Active Vibration Control, Dissertation, KU Leuven, Leuven, Belgium, 1997.
4. P. Gardonio, S. Elliott, Smart panels for active structural acoustic control, Smart Materials and Structures 13 (6) (2004) 1314–1336.
5. S. Elliott, Signal Processing for Active Control, Academic Press, London, UK, 2001.
6. X. Sun, S.M. Kuo, G. Meng, Adaptive algorithm for active control of impulsive noise, Journal of Sound and Vibration 291 (2) (2006) 516–522.
7. R.L. Clark, D.S. Bernstein, Hybrid control: seperation in design, Journal of Sound and Vibration 214 (4) (1998) 784–791.
8. N. Doelman, A unified control strategy for the active reduction of sound and vibration, Journal of Intelligent Material Systems and Structures 2 (1991) 558–580.
9. G. Pavic, Comparison of different strategies of active vibration control, Proceedings of ACTIVE 95, Newport Beach, CA, USA, 1995.
10. J. Pan, C.H. Hansen, Active control of total vibratory power flow in a beam. I: physical system analysis, Journal of the Acoustical Society of America 89 (4) (1991) 200–209.
11. R.L. Clark, C.R. Fuller, Optimal placement of piezoelectric actuators and polyvinylidene fluoride error sensors in active structural acoustic control approaches, Journal of the Acoustical Society of America 92 (3) (1992) 1521–1533.
12. O. Bardou, P. Gardonio, S.J. Elliott, R.J. Pinnington, Active power minimisation and power absorption in a plate with force and moment excitation, Journal of Sound and Vibration 208 (1) (1997) 111–151.

13. M.H.H. Oude Nijhuis, A. de Boer, Optimization strategy for actuator and sensor placement in active structural acoustic control, in: Proceedings of ACTIVE 2002, Southampton, UK, 2002.

14. D.C. Zimmerman, A Darwinian approach to the actuator number and problem placement with non negligible actuator mass, Mechanical Systems and Signal Processing 7 (4) (1993) 363–374.

15. V.V. Varadan, J. Kim, V.K. Varadan, Optimal placement of piezoelectric actuators for active noise control, AIAA Journal 35 (3) (1997) 526–533

16. D.A. Bristow, M. Tharayil, A.G. Alleyne, A survey of iterative learning control, IEEE Control Systems Magazine 26 (3) (2006) 96–114.

17. M. Uchiyama, Formulation of high-speed motion pattern of mechanical arm by trial, Transactions of the Society of Instrumentation and Control Engineers 19 (1978) 706–712.

18. S. Arimoto, S. Kawamura, F. Miyazaki, Bettering operation of robots by learning, Journal of Robotic Systems 1 (1984) 123–140.

19. S. Arimoto, S. Kawamura, F. Miyazaki, Iterative learning control for robot systems, Proceedings of IECON, Tokyo, Japan, 1984.

20. R.W. Longman, Iterative learning control and repetitive control for engineering practice, International Journal of Control 73 (10) (2000) 930–954.

21. R. Horowitz, Learning control of robot manipulators, ASME Journal of Dynamic Systems Measurement and Control 115 (2B) (1993) 402–411.

22. K.L. Moore, Iterative Learning Control for Deterministic Systems, Springer, London, UK, 1993.

23. K.L. Moore, Iterative learning control: an expository overview, Applied and Computational Controls, Signal Processing, and Circuits 1 (1) (1998).

24. P.B. Goldsmith, On the equivalence of causal LTI iterative learning control and feedback control, Automatica 38 (2002) 703–708.

25. G. Pinte, W. Desmet, P. Sas, Active control of repetitive transient noise, Journal of Sound and Vibration 207 (3–5) (2007) 513–526.

26. G. Pinte, W. Desmet, P. Sas, Active structural acoustic control of impact noise, Proceedings of ISMA 2004, Leuven, Belgium, 2004.

27. Z. Bien, K.M. Huh, Higher-order iterative learning control algorithm, IEEE Proceedings Part D, Control Theory and Applications 136 (1989) 105–112.

28. M. Norrlo"f, S. Gunnarsson, A frequency domain analysis of a second order iterative learning control algorithm, in: Proceedings of 38th IEEE Conference on Decision and Control, Phoenix, AZ, USA, 1999.

29. M. Norrlo"f, Iterative Learning Control: Analysis, Design, and Experiments, Dissertation no. 653, Linko"ping Studies in Science and Technology, Sweden, 2000.

30. M.H.A. Verwoerd, Iterative Learning Control—A Critical Review, Dissertation, University of Twente, Enschede, the Netherlands, 2005.

CITATION

G. Pinte, R. Boonen, W. Desmet, P. Sas, Active structural acoustic control of repetitive impact noise, Journal of Sound and Vibration, Volume 319, Issues 3–5, 23 January 2009, Pages 768-794, ISSN 0022-460X, doi.org/10.1016/j.jsv.2008.07.016.

CHAPTER 4

Interacting Cracks Analysis Using Finite Element Method

RuslizamDaud, Ahmad Kamal Ariffin, Shahrum Abdullah and Al Emran Ismail

Universiti Kebangsaan Malaysia, Malaysia

INTRODUCTION

This chapter aims to introduce the concept of fracture mechanics and numerical approaches to solve interacting cracks problems in solid bodies which involves elastic crack interaction. The elastic crack interaction is a result of changes in stress field distribution as the applied force is given during remote loading. The main emphasis is to address the computational evaluation on mechanistic models based on crack tip displacement, stress fields and energy flows for multiple cracks. This chapter start with a brief discussion on fracture and failure that promoted by interacting cracks from industrial cases to bring the issues of how important the crack interaction behaviour is. The present fracture and failure mechanism is assumed to exhibit the brittle fracture. Thus, the concept of linear elastic fracture mechanics (LEFM) is discussed regarding the crack interaction model formulation. As the elastic crack interaction is concerned, the previous analytical and numerical solution of crack interaction are elaborated comprehensively corresponds to fitness-for-service (FFS) as published by ASME boiler and pressure vessel code (Section XI, Articles IWA-3330), JSME fitness-for-service code and BSI PD6493 and BS7910. A new computational fracture mechanics algorithm is developed by adopting stress singularity approach in finite element (FE) formulation. The result of developed approach is discussed based on the crack interaction limit (CIL) aspects and crack unification limit (CUL) in pertinent to the equality of two cracks to single crack rules in FFS. As a conclusion, the FE formulated approach was found to be at agreeable accuracy with analytical formulation and FFS at certain range of crack interval.

Fracture and Failures by Interacting Cracks

This section provides the overview of failure cases in industries and related fitness-for-sevice (FFS) codes which used to assess any cracks or flaws that detected in structures. The works on solution models for FFS codes improvement in specific cases of interacting cracks also discussed.

Industrial Failures

In this section, the fracture and failure by interacting cracks is explained by examples from industrial failures. Mechanical structures and components are designed with multiple stress concentration features (SCF) such as notches, holes, corners and bends. For example, welding and riveting in joining and fastening process have consumed to the increase of stress concentration factor. In every SCF, there is a critical point that experienced the highest concentrated stress field, named as multiple stress riser points (MRSP). Under multiple mode of loading and environmental effect, the interaction between SCF and MRSP tends to form multiple cracks in various types of cracks (e.g. straight crack, surface crack and curved crack) before the cracks propagate in various path to coalescence, overlapping, overlapping, branching and finally fracture in brittle manner.

Crack interaction that induced from MRSP has caused many catastrophic failures, for example in aircraft fuselage (Hu, Liu, & Barter, 2009), F-18 Hornet bulkhead (Andersson et al., 2009), rotor fault (Sekhtar, 2008), gigantic storage tanks (Chang & Lin, 2006), oil tankers (Garwood, 2001), polypropylene tank (Lewis &Weidmann, 2001) and the most recent is the fail of helicopter longerons (J. A. Newman, Baughman, & Wallace, 2010). The recent lab experimental work on multiple crack initiation, propagation and coalescence by (Park &Bobet, 2010) and metallurgical work by (J. A. Newman et al., 2010) supported the important role of crack interaction in fracture and failure. The above cases proved how crucial and important the research on crack interaction is.

To explain further, failure in aircraft is considered as an example. Al Alloy has been extensively used for the fabrication of fuselage, wing, empennage, supporting structures that involve many fastening and joining points. Under static, cyclic loading and environmental effect, the micro-cracks are initiated from MRSP. To certain extent, out of many factors, brittle failure may happen through catastrophic failure (Hu, Liu, & Barter, 2009). As the distance between MRSP is close, the interaction between cracks is become more critical. The fracture behavior due to interacting cracks as the distance between cracks is closed need more understanding. The conventional fracture mechanics may be insufficient to support.

In this case, the advancement of computational fracture mechanic may contribute a lot in crack interaction research and increase the accuracy of failure prediction (Andersson et al., 2009). Most recent structural failure being reported is dealing with fuselage joints in large aircraft structures (Hu et al., 2009). The aircraft fuselage structure is made by 7050-T7451 aluminum alloy and designed to have multiple shallow notches that purposely used for lap joints. Due to environmental reaction and variable magnitude of operational loading, multi-site cracks are formed at MRSP. Therefore, the interaction of multi-site cracks is needed to be quantified for possible coalescence between cracks. The challenge is how to predict the accurate coalescence and fatigue crack growth for the multiple crack problems. Table 1 provides the list of industrial failures that originated from various kinds of multiple cracks. Under loading condition such as mechanical or thermal loading, it is observed that most of the cracks or flaws formation is start on the surface of the body rather than embedded inside the body.

Crack interaction intensity exist in the form of stress shielding and amplification. The failure mechanism by cracks interaction may occurs under brittle, plastic and creep failure. In general, the interacting cracks problems that promotes the fracture and failure of structures are solved using the advancement of fracture mechanics. The fracture mechanics solution can be accomplished through analytical, numerical and experimental work. Based on the above individual approach or combination of them, typically the solution is represented by a model. The model may be defined based on the uncertainty input for the model such as crack geometry, loading and material properties. For crack interaction problems that randomness is relatively small, the deterministic analysis is the best to considered rather than probabilistic analysis. The model is suitable for any deterministic system response. However, when the randomness is relatively high, the model system response required a more robust solution, known as probabilistic approach.

Multiple cracks interaction can be defined into elastic crack interaction and plastic crack interaction that may be referred to theory of elasticity and plasticity, respectively. Under loading condition, the high stresses near crack tips usually accompanied by inelastic deformation and other non-linear effect. If the inelastic deformation and other non-linear effect are relatively small compared to crack sizes and other geometrical body characteristic, the linear theory is most adequate to address the crack interaction behavioral problems. Thus, the role of elastic driving force originated from crack tips can be translated into elastic crack interaction. Then, for every type of interaction, it may classified into interaction without crack propagation (EIWO) where the interaction occurs in the region of SIF $K<K_c$ and interaction with crack propagation (EIWI) occurs in the region of SIF $K \geq K_c$. In this case of EIWO, the quantification of crack extension is neglected. The EIWO becomes the main issue of interaction in present study since it is inadequate investigation on crack

interaction limit and multiple to single crack equivalencies. The study on EIWO is typically measured the fracture parameter and its behavior based on SIF while EIWI more focused to evaluation and prediction of crack path, propagation, coalescence, branching and crack arrestment. Under mechanical or thermal loading condition, the generated interaction will varies depends on type of loading mode (e.g. Mode I and Mode II) being applied. The preceding sections outlined the related fitness-for-service (FFS) codes in pertinent to failures that caused by crack interaction.

Table 1. Summary of industrial failures caused by multiple crack interaction

Multiple Cracks	Industrial Failures	References
Collinear cracks	Failure of crack arresters-stiffeners in aircraft structures	(Isida, 1973)
Parallel and layered cracks	Failure of welded-bonded structures in composite structures for aircraft	(Ratwani & Gupta, 1974)
Collinear cracks and edge cracks	Catastrophic fracture accidents of turbine or generator motor	(Matake & Imai, 1977; Pant, Singh, & Mishra, 2011; Sekhtar, 2008)
Elliptical cracks	Failure in boilers	(O'donoghue, Nishioka, & Atluri, 1984)
Collinear and radial cracks	Failure of pressurized thick-walled cylinder	(Chen & Liu, 1988; Kirkhope, Bell, & Kirkhope, 1991)
Collinear and micro-cracks	Failure of ceramic material in heat exchangers and automobiles	(Lam & Phua, 1991)
Parallel cracks	Failure of aero-engine turbine engines coatings	(Meizoso, Esnaola, & Perez, 1995)
Parallel edge cracks	Failure of actuator piston rods	(Rutti & Wentzel, 1997)
Edge cracks	Brittle failure of Oil Tanker structures at welded joints	(Garwood, 2001)
Array of edge cracks	Failure of heat-checked gun tubes and rapidly cooled pressure vessels	(Parker, 1999)
Semi-elliptical surface cracks	Failure in pressure vessel and piping components	(Moussa, Bell, & Tan, 1999; Murakami & Nasser, 1982)
Collinear cracks and flat elliptical cracks	Multiple site damage in aircraft structures	(Gorbatikh & Kachanov, 2000; Jeong & Brewer, 1995; Jones, Peng, & Pitt, 2002; Milwater, 2010; Pitt, Jones, & Atluri, 1999)
Multiple flaws and surface cracks	Failure of nuclear power plant components	(Kamaya, Miyokawa, & Kikuchi, 2010; Kobayashi & Kashima, 2000)
Penny-shaped cracks	Brittle fracture of welded structures in pressure vessels	(Saha & Ganguly, 2005)
Offset collinear and layered cracks	Fracture and catastrophic failure in polymeric structures	(Lewis & Weidmann, 2001; Sankar & Lesser, 2006; Weidmann & Lewis, 2001)
Interface cracks	Failure in electronic packages and micro-electro-mechanical systems (MEMS)	(Ikeda, Nagai, Yamanaga, & Miyazaki, 2006)
Parallel cracks	Fracture in functional gradient materials (FGM)	(Yang, 2009)
Short cracks and micro-cracks	Fracture in bones	(Lakes, Nakamura, Behiri, & Bonfield, 1990; Mischinski & Mural, 2011; Ural, Zioupos, Buchanan, & Vashishth, 2011)

Fitness-For-Service (FFS) Codes

This section presents the guidelinea that have been published in fitness-for-service (FFS) codes. The related investigation works also discussed as the FFS codes are evaluated in different case of interacting cracks problems. ASME Boiler and Pressure Vessel Code Section XI (ASME, 1998, 2004) and API 579 (ASME, 2007) defines that the multiple cracks are assumed to be independent until or unless the following conditions are satisfied. In the case of two parallel cracks in solid bodies, if distance between crack planes d $d \leq 12.7mm$, the cracks are treated as being coplanar. For coplanar cracks, if the distance between cracks ss$\leq 2 \times$(maximum of$\{a_i\}_{i=1,2}$). Indeed, the single enveloping crack is assumed equal to coplanar cracks if the condition of crack depth a and surface length c satisfy a=maximum of$\{a_i\}_{i=1,2}, c = c_1 + c_2 + s/2$. If the non-coplanar cracks in overlapped condition at s<0, the cracks are assumed to be coplanar, the surface length c cnew=$c_1 + c_2 + s$. Thus, coalescence occur at s=max($2a_1, 2a_1$). In United Kingdom, the engineering critical assessment (ECA) of potential or actual defects in engineering structures is codified into two prime standard; British Standard PD6495 (BSI, 1991) and Nuclear Electric CEGB R6 (R6, 2006). BSI PD6495 has replaced by BS7910 (BSI, 1997), and most latest (BSI, 2005). Original BSI PD6495 primarily concerned on assessment of defect welds. The PD 6495 used crack tip open displacement (CTOD) and SIF.

K based analysis while the BS7910 used CTOD, K or equivalent K that derived from J-integral. Both codes define the cracks are assumed to be independent until or unless the following distance between crack planes d satisfy $d \leq 0.5(a_1 + a_2)$ and the cracks are treated as being coplanar. For coplanar cracks, if the distance between cracks s s$\leq 2 \times$(minimum of$\{c_i\}_{i=1,2}$), a single enveloping crack is assumed and the coalescence occur at s=$0.5(2c_1 + 2c_2)$. The R6 is an approach to upgrade the BSI PD6495 by Central Electricity Generating Board (CEGB) that focused on operating equipment at high temperature where the assumption of equal or greater fracture and possibility of plastic collapse together and fracture separately. The concept of failure assessment diagram (FAD) is introduced to occupy the need of fracture parameter of plasticity fracture. R6 provides a special form of J-integral analysis to impose the plastic collapse limit. Details of FAD can be found in (R6, 2006). JSME Fitness-for-Service Code provides no prescription for the interference between multiple cracks or flaws. In (JSME, 2000), in example of parallel offset cracks, the multiple cracks are replaced by an equivalent single crack based on the stage of detected cracks with satisfying the condition of S$\leq$5mm (H$\leq$10mm) and S>5mm(H<2S) where S is relative vertical spacing and H is relative horizontal spacing. When the crack tips distance S$\leq$0 due to overlapped condition, the crack growth evaluation is considered about the coalescence stages. The guideline in JSME Code is based on experimental results. In (JSME, 2008), the interacting cracks are combined in crack growth prediction and the judgment is based on the relative spacing S and H at the initial condition. If the relative spacing at the beginning of the growth prediction meets the criterion, two cracks are combined when the distance S become zero during the crack growth.

Interacting Cracks Models for Fitness-For-Service

The solution for interacting surface or embedded cracks that based on FFS revision is limited in literature especially for interacting parallel edge cracks in finite body. Therefore, a review on available developed technique or models that related to FFS is presented in this section. The need of continues revision on FFS limitations based on ASME Boiler and Pressure Vessel Code Section XI and British Standard BS7910 started by the industrial failures in all major pressure vessels (Burdekin, 1982). The pressure vessels are designed and built to comply with ASME Boiler and Pressure Vessel Code Section XI and British Standard BS7910 codes but the failure occurrences are significantly high. Both codes can be expressed as

$$
\begin{array}{ll}
\text{ASM} & s \leq 2 \times (\text{maximum of} \{a_1, a_2\}) \\
\text{BSIPD6493} & s \leq 2 \times (\text{maximum of} \{c_1, c_2\})
\end{array}
\tag{1}
$$

To investigate the problem, (Burdekin, 1982) studied the interaction between the collapse and fracture in pressure vessels using the approach that successfully applied to nuclear applications. The approach applied the LEFM and EPFM using COD and J-integral based on single crack under bending condition. The study revealed the important of fracture mechanics as a tool for interaction in failures. This work can be considered as among the first work that put concern on the FFS codes.

Similarly, (O'donoghue et al., 1984) investigated the formation of elliptical cracks in aircraft and pressure vessel attachment lugs and identified the formation is due to stress risers and cracks interaction. Two equal coplanar surface cracks under Mode I loading are modeled using the proposed finite element alternating method (FEAM) in finite solid and the FE analysis results are compared to ASME Boiler and Pressure Vessel Code procedure. The interaction effects are defined by proposed magnification factor (normalized SIF) and the magnification factors seem to increase due to the increase interaction of two cracks and the depth of cracks. The Section XI of ASME Boiler and Pressure Vessel Code recommend that two interacting surface flaws in a pressure vessel should be modeled by a single elliptical crack that covers both flaws. It can be seen that SIF for single crack as proposed by FFS code are generally larger than those due to two interacting cracks as proposed by (O'donoghue et al., 1984). This trend of magnification factor shows that the ASME Boiler and Pressure Vessel Code in Section XI procedure will tend to underestimate the design life of multiple flaws structures. The ASME pressure vessel codes (ASME, 1998) and British Standard PD6495 (BSI, 1991) do not quantify the interaction between cracks especially in two close proximity cracks. At sufficiently close distance, the interaction may cause the increase of SIF. The exclusion of crack interaction may result with unrealistic SIF. Therefore, with the concern on the above standard guideline, (Leek & Howard, 1994) presents an empirical method to approximate the interaction factor of two coplanar surface cracks under tension and bending loading. The approximation approach resulted with good agreement with FE analysis using developed BERSAFE program, (Murakami & Nasser, 1982) and (J. C. Newman &Raju, 1981) within ± 5% discrepancy.

$$a=\text{maximum of}\{a^i\}_{i=1,2}, c=c_1+c_2 \tag{2}$$
$$(s/c^-)\times(s/a^-)>3.38 \quad \text{and} \quad s/a^- \geq 2.49 \tag{3}$$

Based on the (ASME, 2004) and (BSI, 1991) design code that expressed by Eq. (1), (Moussa et al., 1999) used FEM to analyze interaction of two identical parallel non-coplanar surface cracks subjected to remote tension and pure bending loads. The interaction factor as a function of stress shielding to cause overlapping in distance is studied using three dimensional linear finite element analyses. The formation of stress relaxation state is introduced near crack front, as a form of shielding effect at sufficient overlapping. J-integral is calculated based on models by (Shivakumar&Raju, 1992) and the interaction factor γ is defined as follows:

$$\gamma=K_{in}/K_{is} \tag{4}$$

where K_{in} and K_{is} are the SIFs with and without the influence of interaction, respectively. As conclusion, the interaction effect appears to diminish as the value of s/c approaches 2.0. The existing rules for re-characterization of interacting cracks as less conservative for high values of s/c and over-conservative as s/c is close to 2.0.

The existing (ASME, 1998; JSME, 2000) FFS combination rules provided no prescription for the interference between multiple cracks for corrosion fatigue. Therefore, (Kamaya, 2003) developed simulation model to extent the condition of coalescence rules in (JSME, 2000) for crack growth process using body force method (BFM). BFM is used to investigate the multiple cracks growth in stress corrosion cracking. Based on JSME code and the SIF value of coalescence behavior from experiments, the new SIF formulation is developed using BFM where focus is given to the interaction between cracks under various relative position and size. The crack propagation direction can be written as

$$\theta_{max} = \mp \cos^{-1}\left(\left(3K_{II}^2 + K_I\sqrt{8K_{II}^2 + K_I^2}\right)/\left(9K_{II}^2 + K_I^2\right)\right)$$

(5)

where the sign in Eq. (5) positive in the case of $K_{II}/K_I < 0$ and negative in the case of $K_{II}/K_I > 0$. When the crack are close and overlapped, the crack interaction intensity between cracks is almost equivalent to single coalesced crack. The change of inner crack tips direction also found with little influence on the crack growth behavior. The relative crack length and position influenced the crack interaction intensity.

The combination rule in ASME Code is found to provide the relative large overestimation of the actual crack growth since the complex growth phenomena under interaction are summarized in simple combination rules. In order to reduce the conservativeness in existing code, (Kamaya, 2008b) proposed alternative assessment procedures based on the size of area and fatigue crack growth. Experimental analysis and testing is conducted using stainless steel specimens (A-H0S5 and B-H0S5) subjected to cyclic tensile loading. FE analysis is carried out to simulate the crack growth during coalescence. In the simulation, the automatic meshing was generated by command language in PATRAN and the SIF is derived from energy release rate obtained from virtual crack extension integral method using ABAQUS. The normalized SIF of Mode I, K_I is expresses as

$$F_I = K_I / \sigma_0\sqrt{\pi a}$$

(6)

Where σ_0 denotes the applied tensile stress and the a is the maximum depth. As a result, the area of the crack face is concluded to be the predominant parameter for the crack growth of interacting cracks under test condition. The cracks of various shapes can be characterized as semi-elliptical cracks of the same area. In extension of parallel semi-elliptical cracks study, (Kamaya, 2008a) investigated the coalescence of adjacent cracks as a result of crack growth with the influence of crack interaction. The magnitude of interaction is represented by driving force of the crack growth (CGF), written as

$$W_m = \sum_{i=1}^{n-1} 0.5\left(D_p K_{I(i)}^{m_p} + D_p K_{I(i+1)}^{m_p} \right)\Delta g$$

(7)

whereDp and m_p are the material constant, $K_I(i)$ denotes the Mode I SIF of the ith node fromp=0°. Δg and n are the distance of neighboring mode on the crack front and number of nodes. The CGF formulation proved that the interaction between surface cracks not only dependent on relative spacing but also the position of crack front. In the condition of S > 0, as the cracks overlapped, the stress shielding effect influenced the change of CGF. The most important, the study notified the cracks can be replaced with single crack of the same area when the relative spacing is sufficiently close, at crack spacing H < a. In regular inspection of pressure vessel components, the adjacent defects are found close enough. Under operational loading, the stress field around the crack tips will be magnified and accelerates the crack growth rate. This matter has been referred to current fitness-for-service (FFS) rules such as ASME Boiler and Pressure Vessel Code Section XI (ASME, 1998, 2004), API 579 (ASME, 2007), British Standard PD6495 (BSI, 1991), BS7910 (BSI, 1997, 2005) and Nuclear Electric CEGB R6 (R6, 2006). The multiple interacting cracks are combined as single crack as the two cracks satisfy the prescribed criterion. As observed, this rule introduced unrealistic discontinuity in the process of crack growth due to the crack interaction is neglected. The evaluation of two interacting coplanar cracks in plates under tension is conducted by (Xuan, Si, &Tu, 2009) and creep interaction factorγ_{creep} is introduced by using C*integral prediction analysis and the FE analysis is executed using ABAQUS to verify the proposed approach. Creep interaction factorγ_{creep} is expressed as

$$\gamma_{creep} = \left(C_{Double}^* / C_{Single}^* \right)^{1/2}$$

(8)

In conclusion, the creep crack interaction represented by C*integral is affected by crack configuration (e.g. relative crack distance c/d, depth of crack a/t and location at crack front $2\o/\pi$) and time dependent properties of material such as creep exponent n. The increasing crack aspect a/c resulted with no significant effect to C* integral.

Most recent, (Kamaya et al., 2010) used S-version finite element method to determine the SIF changes due to the interaction of stress field which caused variation in crack growth rate and cracks shape. The root of interaction problems is referred to (JSME, 2000, 2008) and (ASME, 2004, 2007) for the case of interacting dissimilar crack sizes. However, the effect of difference crack size or relative size effect is not taken into account in the aforementioned code. The results have shown that smaller cracks stopped growing when the difference in size of interaction was large enough. It means, the interaction effect on the fatigue life of the larger cracks was negligibly small. Moreover, the offset distance and the relative size were important parameter for interaction evaluation especially when the S=0 and the condition of crack spacing H/c_1 and cracks ratio c_2/c_1 must be considered most. In present study, the focus is given to determine the stage of crack interaction intensity is equal to single crack in a state of crack interaction limit (CIL) and unification (CUL) using finite element method.

FINITE ELEMENT ANALYSIS

The stress in the neighborhood of a crack tip in homogenous isotropic material exist in a form of square-root singular and there have been many special elements or singularity function based approach were described in details in (Banks-Sills, 1991). The square root singular stresses in the neighborhood can be modeled by quarter-point, square and collapsed, triangular elements for two dimensional problems, and by brick and collapsed, prismatic elements in three dimensions. Quarter-points square have been found to produce the most excellent results (Banks-Sills, 2010). The stiffness matrix of the element is evaluated using two-dimensional integral based on Gaussian quadrature approach. The plate is constructed with a consideration of singular element and assigned to both crack tips Ct1 and Ct2. It is because the high gradients of singular stress-strain and deformation fields are concentrated at both crack tips. The SIF calculation is limited to linear elastic problem with a homogeneous, isotropic material near the crack region.

Singularity Stress Field
The studies are conducted in a pure Mode I loading condition with specified material, Alloy 7475 T7351 solid plate in constant thickness, homogenous isotropic continuum material, linear elastic behavior, small strain and displacements, and crack surface are smooth. According to Westergaard method for single crack, Mode I K_I and Mode II K_{II} SIF can be expressed as:

$$K_I = F / K_o = F(b/a, a/W)\, \sigma_o \sqrt{\pi a}$$
$$K_{II} = F / K_o = F(b/a, a/W)\, \tau_o \sqrt{\pi a}$$

$$(9)$$

whereσ_o is nominal stress, $\tau_o$is shear stress, W is width of specimen, a and b is the length of crack and crack interval, respectively. The work starts by determination of K_I and K_{II} using Eq. (9). The important issue which differs from single crack is the existence of cracks interaction in fracture analysis.

Consider two multiple edge crack of length a1and a2 which occupies the segment of $0.05 \leq a/W \leq 0.5$ and $0.5 \leq b/a \leq 3.0$ in finite plate subjected to uniform equal stress σ along the y direction, as shown in Fig. 1(a) and (b). The SIF formulation is based on the creation of singular element at the crack tip based on quadratic isoparametric finite element developed in ANSYS evironment based on (Madenci&Guven, 2006), where the element is based on Barsoum (Barsoum, 1974, 1975), as depicted in Fig. 2 (a) and (b). The singularity is obtained by shifting the mid-side node the ¼ point close to the crack tip. To calculate the SIF, the elements are assumed to be in rigid body motion and constant strain modes. The master element mapping in Cartesian space is transformed into curvilinear space using Jacobian transformation which used to interpolate the displacement within the elements (Chandrupatla&Belegundu, 2002). The accuracy of special element has been addressed by (Murakami, 1976) where the crack tip nodal point is enclosed by a number of special element. In analysis, the size, number and compatibility of special elements really affect the accuracy. The special elements also defined as singularity function methods where stress singularity at crack tip is modeled. The condition of continuity between elements is the most important. By using singularity function method, Mode I and Mode II of stress intensity factor may be able to calculate with high accuracy (Shields, Srivatsan, &Padovan, 1992).

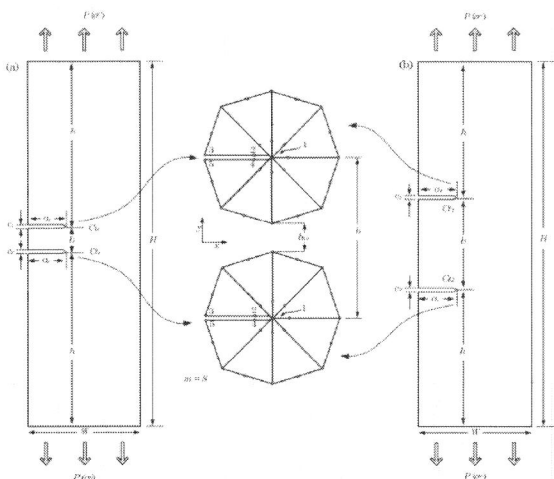

Figure 1. Barsoum singular element for (a) strong crack interaction and (b) weak crack interaction.

According to Eq. (9), the shape correction factor can be converted to new elastic interaction factor

$$\gamma_{I,in,D} = K_I / K_o \tag{10}$$

$$\gamma_{II,in,D} = K_{II} / K_o \tag{11}$$

whereγ_I and γ_{II} denotes to elastic interaction factor for Mode I and II fracture, respectively.

The SIF for Mode I and Mode II is determined using Displacement Extrapolation Method (DEM) by written APDL macro code in ANSYS (Madenci&Guven, 2006), and expressed as

$$K_I = \frac{E}{3(1+v)(1+\kappa)} \sqrt{\frac{2\pi}{L}} \left(4(v_2 - v_4) - \frac{(v_3 - v_5)}{2} \right) \tag{12}$$

$$K_{II} = \frac{E}{3(1+v)(1+\kappa)} \sqrt{\frac{2\pi}{L}} \left(4(u_2 - u_4) - \frac{(u_3 - u_5)}{2} \right) \tag{13}$$

where, E=Young Modulus, $\kappa=3-4v$ for plain stress, $\kappa=3-4v/1-v$ for plain strain, L is length of element, v and u are displacements in a local Cartesian coordinate system and υ is Poisson's ratio.

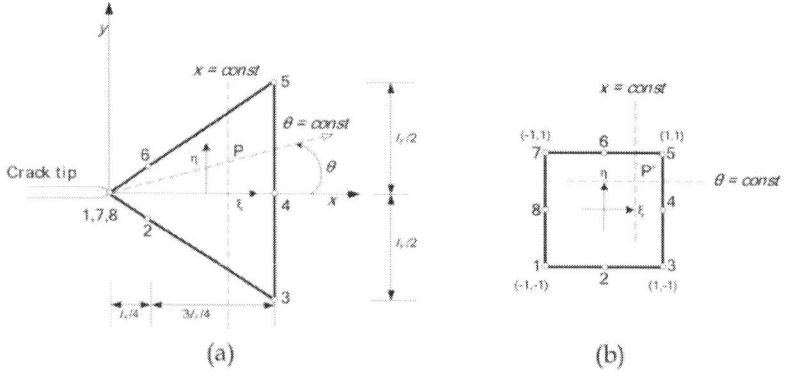

Source: (Barsoum, 1974, 1976; Henshell & Shaw, 1975)

Figure 2. (a) Eight nodes quadratic isoparametric elements (b) Parent element.

Findings and discussion

Two parallel edge cracks interaction will mainly referred to shielding effect rather than amplification effect. The crack interaction is proportional to the magnitude of elastic interaction factor γ_I. The crack interaction will only exist at b/a < 3 (Z.D.Jiang, A.Zeghloul, G.Bezine, &J.Petit, 1990; Z.D.Jiang, J.Petit, &G.Bezine, 1990), the analytical formulation can be expressed as

$$K_I = K_o F_n (a / W, b / a)$$

$$(14)$$

In which

$$\gamma_{I,in,Ji} = F_n = F_2 = \frac{A_0 + A_1 (a/w)^{1.5} + A_2 (a/w)^4}{\sqrt{1 - (a/w)^2}}$$

$$A_0 = 0.79 + 0.07 (b/a) + 0.04 (b/a)^2 - 0.011 (b/a)^3$$

$$A_1 = 1.74 + 2.84 (b/a) - 1.44 (b/a)^2 + 0.206 (b/a)^3$$

$$A_2 = 6.02 + 2.19 (b/a) - 3.26 (b/a)^2 + 0.828 (b/a)^3$$

$$(15)$$

And analytical single edge crack SIF reference by (Brown &Strawley, 1966),

$$K_{I,ref} = \sigma_0 \sqrt{\pi a} \left(1.12 - 0.23 (a/W) + 10.6 (a/W)^2 - 21.7 (a/W)^3 + 30.4 (a/W)^4 \right)$$

$$K_{I,ref} = \sigma_0 \sqrt{\pi a} \left(f_{I,ref,BS} \right)$$

$$(16)$$

Crack Interaction Factor $\gamma_{i,In,D}$ Comparison With Analytical Data $\gamma_{i,In,Ji}$

The mode I fracture of the elastic interaction factors $\gamma_{I,in,D}$(I,in,D,ct1,$\gamma_{I,in,D,ct2}$) for crack interval ratio b/a=1.5−3.0 and a/W=0.05−0.5 are shown in Fig. 3-6. Overall, it can be seen that the interaction factor varies with the different a/W values where the crack interaction factor increases as the a/W increases and vice versa. The point of intersection also observed occurred for all the b/a values. For example, the plot of $\gamma_{I,in,D}$($\gamma_{I,in,D,ct1}$,$\gamma_{I,in,D,ct2}$) against a/W at b/a=3.0 are illustrated in Fig. 3.

The $\gamma_{I,in,D}$ prediction line is compared with the predicted result of single edge crack $f_{I,ref,BS}$. A general good agreement can be observed with a minimum difference 0.6% at a/W=0.1 and maximum difference 5.39 % differencea/W=0.15. In comparison, the present $\gamma_{I,in,D}$ has demonstrated more accurate prediction compared with $\gamma_{I,in,Ji}$results. For example, in reference to$f_{I,ref,BS}$, at a/W=0.5, the present prediction $\gamma_{I,in,D}$ is in difference of $\gamma_{I,in,D,ct1}$0.85 % and $\gamma_{I,in,D,ct2}$0.4%, while $\gamma_{I,in,Ji}$ is at 2.15% difference. In terms of the CIL point, the closer the crack interaction to $f_{I,ref,BS}$ of single crack, a more accurate CIL prediction can be achieved. It is noted that the$\gamma_{I,in,Ji}$analytical expression was formulated using the numerical results of J-integral analysis. The formulation is unable to calculate the crack interaction factor for both crack tips and become the weakness of $\gamma_{I,in,Ji}$. Therefore, the present work of DEM has improved the existing J-integral analysis by improving the accuracy of CIL to predict fracture due to crack interaction.

Theoretically, the study of intersection point is most significant in identification of crack interaction limit (CIL) and crack unification limit (CUL). The intersection point of two cracks $\gamma_{I,in,D}$with single crack $f_{I,ref,BS}$ justifies the realization of CIL at higher a/W and CUL at lower a/W. The CIL and CUL also differ with different b/a. From Fig. 3, the identified CUL is at a/W=0.1 and CIL approximately at a/W=0.5.

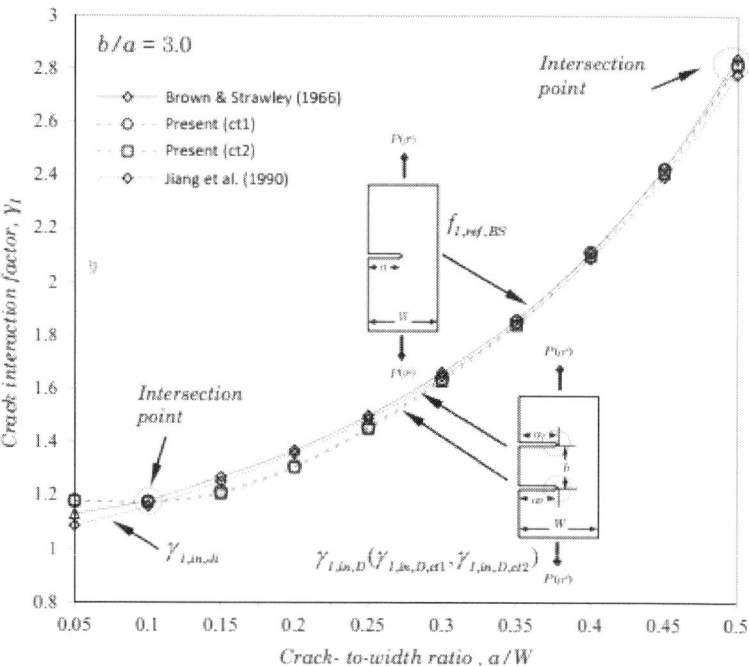

Figure 3. Variation of γI, in, D against a/W for b/a=3.0.

Another significant improvement of present γI,in,D is the moving intersection point as a/Wdecreases for every b/a, as shown in Fig. 3-6. From Fig. 4, the moving intersection point can be noticed moves from a/W=0.1 to a/W=0.075. The moving intersection point exhibits similar prediction trend of γI,in,D. The intersection point also can be denoted as the crack unification limit (CUL) point, which indicates the starting point of strong interaction region start approximately at a/W<0.07.

The intersection point is also observed to move from a/W=0.075 for b/a=2.5, a/W=0.06for b/a=2.5, a/W=0.05 for b/a=2.0 and a/W<0.05 for b/a=1.5, as shown in Fig. 4-6. It means that the CUL is not in a fix limit, it exist in dynamic condition which depends on crack interval ratio b/a. Conversely, the γI,in,Ji prediction model overruled the FFS codes because it does not lead to a single independent or combined crack because of not having any intersection point. The intersection point could not be defined by γI,in,Ji prediction model.

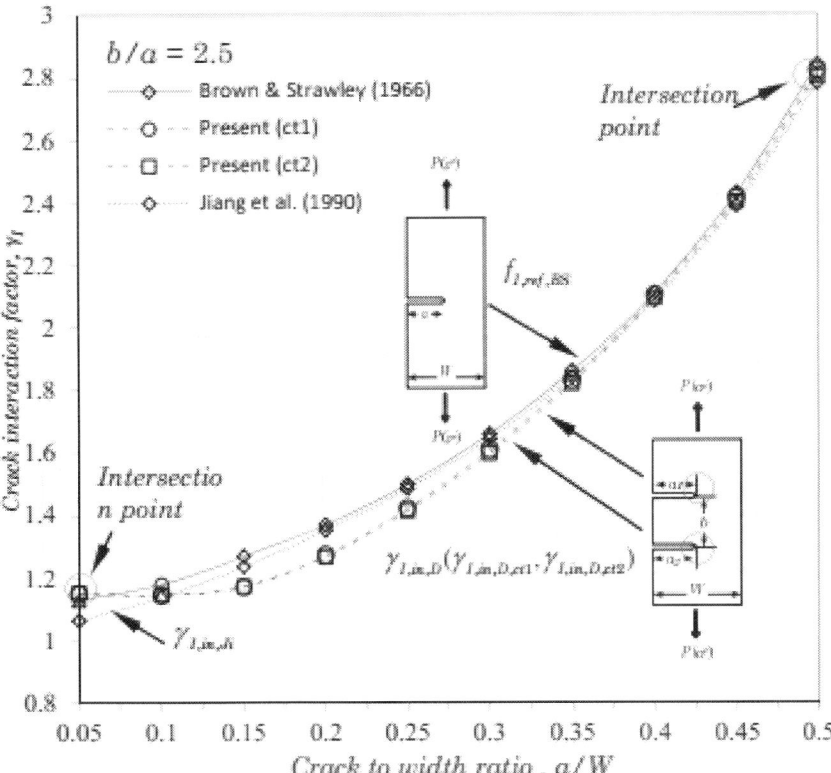

Figure 4. Variation of γI,in,Dagainsta/W for b/a=2.5

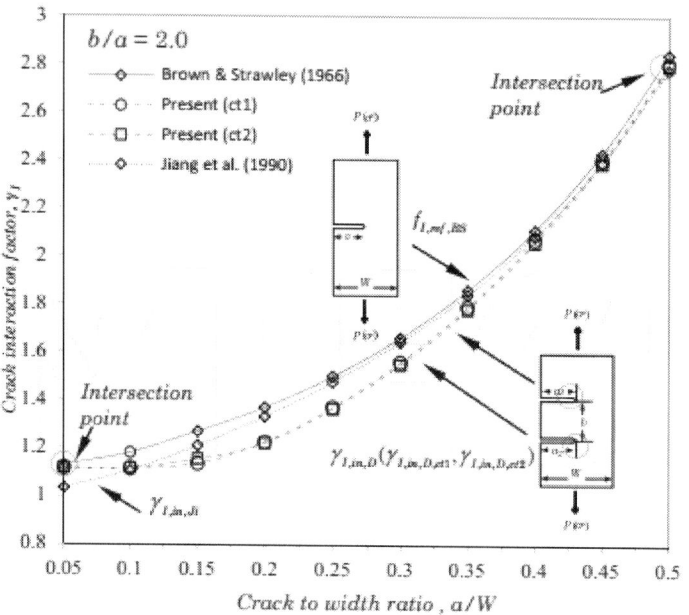

Figure 5. Variation of γI,in,Dagainsta/W for b/a=2.0.

It also can be seen that γI,in,Ji prediction is unable to display the unification of crack interaction factor, which defined equivalent to fl,ref,BS of single crack. In this case, the γI,in,Ji prediction is expected to encounter some numerical errors since at lower a/W<0.07. In analysis, the ratio of crack length and width also define the critical stress field, if the path independent radius of J-integral for both crack tips are overlapped, the calculation of J value might be overestimated and the stress behavior is equal to behavior of single edge crack in finite body. The path integral line should be always apart and controlled in individual condition.

Based on the FFS codes, the multiple cracks are assumed to be independent as single cracks or combined cracks, until or unless certain conditions are satisfied. The intersection point that lies in the present γI,in,D prediction trend curve, which intersects with the single crack prediction, shows good agreement with the outlined FFS codes. It is seen that at range of 0.05≤a/W≤0.15, the γI,in,D value is about to level at 1.072 -1.085. The small changes in these range indicate that shielding effect is very small and promote the unification in interaction at the point of a/W=0.15. The smaller the b/a, the faster unification process starts.

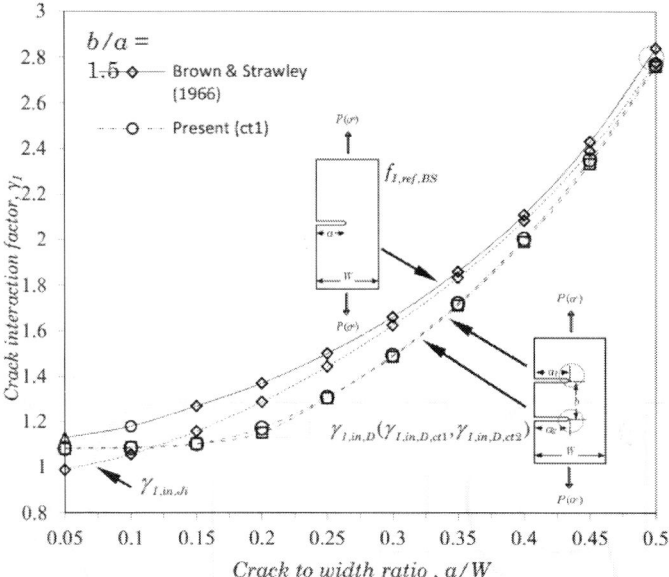

Figure 6. Variation of γI,in,D against a/W for b/a=1.5.

Mode II Fracture Behavior

Fig. 7 depicts the trend of KI,in,D and KII,in,D against b/a for a/W=0.25−0.5. It can be seen that the increase in b/a results in an increase of KI,in,D and a decrease of KII,in,Dvalues. In strong interaction region 0≤b/a≤1.0, the values of KI,in,D increased rapidly for higher crack-to-width ratio a/W=0.5,0.45 and before grew slowly as the value of a/Wdecreases to a/W=0.25.

Meanwhile, the values of KII,in,D declined significantly at higher crack-to-width ratio a/W=0.5,0.45 and before decreased slightly as the value of a/Wdecreases to a/W=0.25. For both crack interaction phase, the value of KI,in,D is always much higher than KII,in,D. In weak interaction region 1.5≤b/a≤3.0, it can be seen that the values ofKI,in,D increased slightly before growing slowly and then maintaining at the same level to steady state at b/a=3.0. At the same region, the value of KII,in,D declined moderately before decreased slightly and remain stable at the level of b/a=3.0. It means that mode II SIF is less influenced by damage shielding effect than mode I SIF. It also defined that the crack opening is more affected by damage shielding effect than the crack sliding. This has been clearly indicated in Fig. 7 (a)-(f) and Fig. 8(a)-(d).

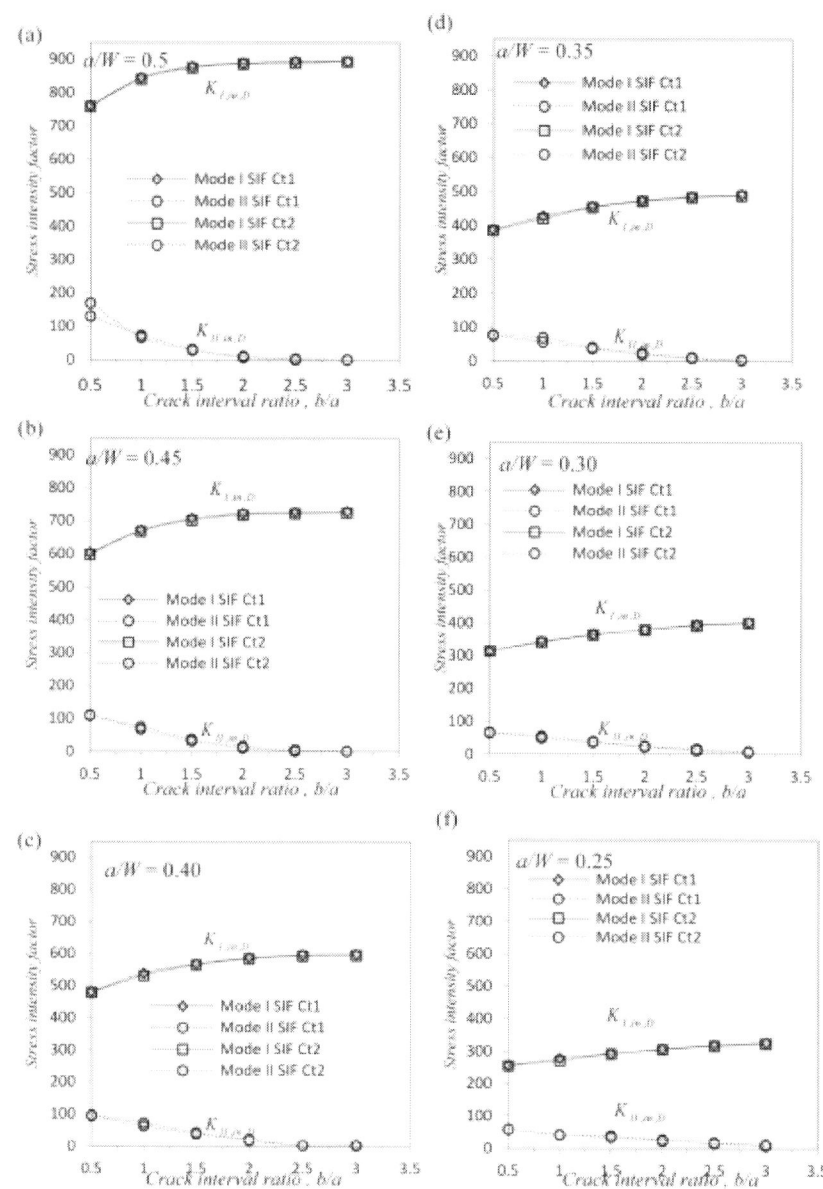

Figure 7. Variation of KI,in,D and KII,in,D for (a/W=0.25−0.5).

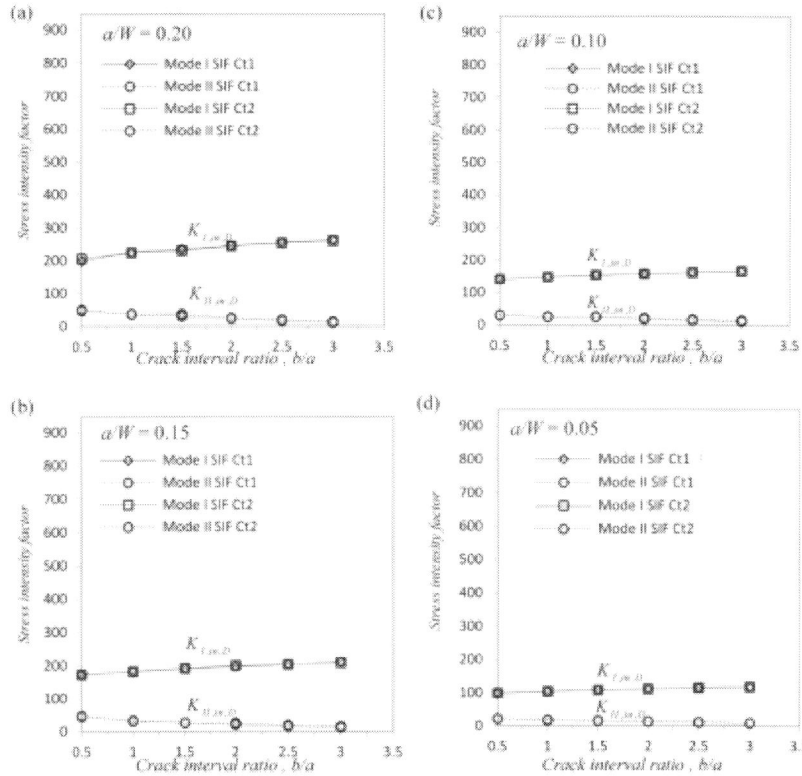

Figure 8. Variation of KI,in,D and KII,in,D for (a/W=0.05−0.2).

From these figures, it also observed that the different between mode II SIF and mode I SIF is reduced as the a/W decreases. In the context of crack interaction limit (CIL) based on KI,in,D and KII,in,D, by considering the convergence level as the indicator of CIL. It can be seen that the degree and speed of CIL achievement not only depends heavily on the increased ofb/a, the reduction of a/W from a/W=0.5−0.05 also provides significant impact on CIL determination.Fig. 8 shows the variation of KI,in,D and KII,in,D for (a/W=0.05−0.2). It can be observed that at smallest a/W=0.05, the value of KI,in,D and KII,in,D is almost hold at constant value and stabilize.

The identification of CIL in this condition is absent because the value of KI,in,D and KII,in,D are about the same value for all b/a. This equalization condition may be referred to the identification of crack unification limit (CIL). Overall, it can be concluded that the higher ratio of b/a and a/W, the more the realization of CIL. Inversely, the lower ratio of b/a and a/W, the more indication to CUL can be realized.

CONCLUSION

The numerical solution based on displacement extrapolation method (DEM) has proved to be more consistent in SIF prediction comparedd to for both crack tips. However, the DEM able to predict the SIF for Mode I and Mode II fracture behaviour. The FE results conclude that the interaction of two cracks is directly influence the reduction of SIF magnitude and γI at the crack tips. The parallel cracks have experienced decrease shielding effect as the cracks interval b/a decrease. The identification of crack interaction limit (CIL) and crack unification limit (CUL) has been accomplished. The SIF of Mode I is found more significant compared to Mode II in higher or lower b/a and a/W ratio. Mode II SIF can be neglected due to its small effect to the stress shielding effect. The FFS codes rules that define the combination of two cracks as single crack are well translated as CIL and CUL.

REFERENCES

1. B Andersson, A. F Blom, U Falk, G. S Wang, K Koski, A Siljander, et al2009A case study of multiple site fatigue crack growth in the F-18 Hornet bulkhead. Paper presented at the 25th ICAF Symposium.
2. ASME. (1998). ASME Boiler and Pressure Vessel Code, Section XI. New York, USA.
3. ASME. (2004). ASME Boiler and Pressure Vessel Code, Section XI. New York, USA.
4. ASME20075791FFS-1 Fitness-for-service, Section 9, American Society of Mechanical EngineersNew York, USA.
5. L Banks-sills, 1991Application of the Finite Element Method to Linear Elastic Fracture Mechanics. Appl. Mech. Rev, 44447461
6. L Banks-sills, 2010Update: Application of the Finite Element Method to Linear Elastic Fracture MechanicsAppl. Mech. Rev, 63117
7. R. S Barsoum, 1974Application of Quadratic Isoparametric Finite Elements in Linear Fracture MechanicsInternational Journal of Fracture10603605
8. R. S Barsoum, 1975Further application of quadratic isoparametric finite element to linear fracture mechanics of plate bending and general rules. International Journal of Fracture, 11167169
9. R. S Barsoum, 1976Application of Triangular Quarter-Point Elements as Crack Tip Elements of Power Law Hardening Material. International Journal of Fracture, 12463466
10. W. F Brown, J. E Strawley, 1966ASTM STP 410.

11. BSI1991British Standard Institute, PD 6493, Section 8. Guidance on methods for assessing the acceptability of flaws in fusion welded structures.

12. BSI1997British Standard Institute, BS7910, Section 8. Guidance on methods for assessing the acceptability of flaws in structures.

13. BSI2005British Standard Institute, BS7910. Guidance on methods for assessing the acceptability of flaws in metallic structures.

14. F. M Burdekin, 1982The role of fracture mechanics in the safety analysis of pressure vesselsInt. J. Mech. Sci., 24(4), 197 EOF208 EOF

15. T. R Chandrupatla, A. D Belegundu, 2002Introduction to Finite Elements in Engineering (3rd ed.): Prentice Hall.

16. J. I Chang, C Lin, C. (2006A study of storage tank accidentsJournal of Loss Prevention in the Process Industries195159

17. Y. Z Chen, H. Y Liu, 1988Multiple cracks in pressurized hollow cyclinder. Theoritical and Applied Fracture Mechanics, 10213218

18. S. J Garwood, 2001Investigation of the MV Kurdistan casuality. Engineering Failure Analysis, 4(1), 3-24.

19. L Gorbatikh, M Kachanov, 2000A simple technique for constructing the full stress and displacement fields in elastic plates with multiple cracksEngineering Fracture Mechanics665163

20. R. D Henshell, K. G Shaw, 1975Crack tip finite elements are unnecessaryInt J NumerEngng, 9495507

21. W Hu, Q Liu, S Barter, 2009A study of interaction and coalescence of micro surface fatigue cracks in aluminium 7050. Paper presented at the 25th ICAF Symposium.

22. T Ikeda, M Nagai, K Yamanaga, N Miyazaki, 2006Stress intensity factor analyses of interface cracks between dissimilar anisotropic materials using the finite element methodEngineering Fracture Mechanics7320672079

23. M Isida, 1973Analysis of stress intensity factors for the tension of a centrally cracked strip with stiffened edgesEngineering Fracture Mechanics5647665

24. D. Y Jeong, J. C Brewer, 1995On the linkup of multiple cracksEngineering Fracture Mechanics233 EOF238 EOF

25. R Jones, D Peng, S Pitt, 2002Assessment of multiple flat elliptical cracks with interactionsTheoretical and Applied Fracture Mechanics, 38281291ssss.

26. JSME2000JSME Fitness-for-Service Code S NA12000

27. JSME2008JSME Fitness-for-Service Code S NA12008

28. M Kamaya, 2003A crack growth evaluation method for interacting multiple cracksJSME International Journal, 46(1), 15 EOF23 EOF

29. M Kamaya, 2008aGrowth evaluation of multiple interacting surface cracks. Part II: Growth evaluation of parallel cracksEngineering Fracture Mechanics7513501366

30. M Kamaya, 2008bGrowth evaluation of multiple interacting surface cracks.Part I: Experiments and simulation of coalesced crack.Engineering Fracture Mechanics7513361349

31. M Kamaya, E Miyokawa, M Kikuchi, 2010Growth prediction of two interacting surface cracks of dissimilar sizesEngineering Fracture Mechanics7731203131

32. K. J Kirkhope, R Bell, J Kirkhope, 1991Stress intensity factors for single and multiple semi-elliptical surface cracks in pressurized thick-walled cylindersInternational Journal of Pressure Vessel and Piping, 47247257

33. H Kobayashi, K Kashima, 2000Overview of JSME flaw evaluation code for nuclear power plantsInternational Journal of Pressure Vessels and Piping, 77937944

34. R. S Lakes, S Nakamura, J. C Behiri, W Bonfield, 1990Fracture mechanics of bone with short cracks.J. Biomechanics, 23(10), 967 EOF75 EOF

35. K. Y Lam, S. P Phua, 1991Multiple crack interaction and its effect on stress intensity factorEngineering Fracture Mechanics585 EOF592 EOF

36. T. H Leek, I. C Howard, 1994Estimating the elastic interaction factor of two coplanar surface cracks under Mode I load.International Journal of Pressure Vessels and Piping, 60307321

37. T. H Leek, I. C Howard, 1996An examination of methods of assessing interacting surface cracks by comparison with experimental dataInternational Journal of Pressure Vessels and Piping, 68181201

38. P. R Lewis, G. W Weidmann, 2001Catastropic failure of a polypropylene tank Part I: primary investigation. Engineering Failure Analysis, 6(4), 197-214.

39. E Madenci, I Guven, 2006The finite element method and application in engineering using ANSYS (1 ed.): Springer Science-Business.

40. T Matake, Y Imai, 1977Pop-in behaviour induced by interaction of cracks. Engineering Fracture Mechanics, 91724

41. AM Meizoso, J. M. M Esnaola, M. F Perez, 1995Interaction effect of multiple crack growth on fatigue. Theoretical and Applied Fracture Mechanics, 23219233

42. H. R Milwater, 2010A Simple, and accurate method for computing stress intensity factors of collinear interacting cracks. Aerospace Science and Technology.

43. S Mischinski, A Mural, 2011Finite Element Modeling of Microcrack Growth in Cortical BoneJournal of Applied Mechanics7819

44. W. A Moussa, R Bell, C. L Tan, 1999The interaction of two parallel non-coplanar identical surface cracks under tension and bendingInternational Journal of Pressure Vessels and Piping, 76135145

45. Y Murakami, 1976A simple procedure for the accurate determination of stress intensity factors by finite element method Engineering Fracture Mechanics8643655

46. Y Murakami, S. N Nasser, 1982Interacting dissimilar semi-elliptical surface flaws under tension and bending Engineering Fracture Mechanics373 EOF386 EOF

47. J. A Newman, J. M Baughman, T. A Wallace, 2010Investigation of cracks found in helicopter longerons Engineering Failure Analysis17416430

48. J. C Newman, I. S Raju, 1981An empirical stress intensity factor equation for the surface cracks. Engineering Fracture Mechanics, 15185192

49. Donoghue, P. E Nishioka, T Atluri, S. N. (1984Multiple surface cracks in pressure vessel. Engineering Fracture Mechanics545 EOF560 EOF

50. M Pant, I. V Singh, B. K Mishra, 2011A numerical study of crack interactions under thermo-mechanical load using EFGMJournal of Mechanical Science and Technology403 EOF413 EOF

51. C. H Park, A Bobet, 2010Crack initiation, propagation and coalescence from frictional flaws in uniaxial compression. Engineering Fracture Mechanics, 7727272748

52. P Parker, 1999Stability of arrays of multiple edge cracksEngineering Fracture Mechanics62577591

53. S Pitt, R Jones, S. N Atluri, 1999Further studies into interacting 3D cracksComputers and Structures, 70583597

54. R6. (2006). Nuclear Electric : Assessment of the integrity of structures containing defects, Revision 4. Gloucester, British Energy Generation Ltd.

55. M Ratwani, G. D Gupta, 1974Interaction between parallel cracks in layered composites. International Journal of Solids and Structures, 10701708

56. T. F Rutti, E. J Wentzel, 1997Investigation of failed actuator piston rods. Engineering Failure Analysis, 5(2), 91-98.

57. T. K Saha, S Ganguly, 2005Interaction of penny-shaped cracks with an eliptic crack under shear loading. International Journal of Fracture, 131267287

58. R Sankar, A. J Lesser, 2006Generic Overlapping Cracks in Polymers: Modeling of InteractionInternational Journal of Fracture, 142277287

59. S Sekhtar, 2008Multiple cracks effects and identification. Mechanical Systems and Signal Processing, 22845878

60. E. B Shields, T. S Srivatsan, J Padovan, 1 EOF26 EOF

61. K. N Shivakumar, I. S Raju, 1992An equivalent domain integral method for three-dimensional mixed-mode fracture problemsEngineering Fracture Mechanics, 42935959

62. A Ural, P Zioupos, D Buchanan, D Vashishth, 2011The effect of strain rate on fracture toughness of human cortical bone: A finite element studyJournal of the Mechanical Behaviour of Biomedical Materials.

63. G. W Weidmann, P. R Lewis, 2001Catastropic failure of a polypropylene tank Part II: comparison of the DVS 2205 code of practice and the design of the failed tank. Engineering Failure Analysis, 6(4), 215-232.

64. F Xuan, Z Si, J Tu, S. T. (2009Evaluation of C* integral for interacting cracks in plates under tensionEngineering Fracture Mechanics7621922201

65. Y. H Yang, 2009Multiple parallel symmetric mode III cracks in a functionally graded material planeJournal of Solid Mechanics and Materials Engineering819 EOF830 EOF

66. Z. D Jiang, A Zeghloul, G Bezine, J Petit, 1990Stress intensity factor of parallel cracks in a finite width sheet.Engineering Fracture Mechanics1073 EOF1079 EOF

67. Z. D Jiang, J Petit, G Bezine, 1990Fatigue propagation of two parallel cracksEngineering Fracture Mechanics1139 EOF1144 EOF

CITATION

RuslizamDaud, Ahmad Kamal Ariffin, Shahrum Abdullah and Al Emran Ismail (2012). Interacting Cracks Analysis Using Finite Element Method, Applied Fracture Mechanics, Dr. Alexander Belov (Ed.), ISBN: 978-953-51-0897-9, InTech, DOI: 10.5772/54358.

CHAPTER 5

Avoidance of Speckle Noise in Laser Vibrometry by the Use of Kurtosis Ratio: Application to Mechanical Fault Diagnostics

J. Vassa,, R. Šmíd[1], R.B. Randall[2], P. Sovka[1], C. Cristalli[3], B. Torcianti[3],

[1] Faculty of Electrical Engineering, Czech Technical University, Technická 2, 166 27 Prague, Czech Republic
[2] School of Mechanical and Manufacturing Engineering, The University of New South Wales, Sydney 2052, Australia
[3] AEA s.r.l., The Loccioni Group, Via Fiume 16, 60030 Angeli di Rosora (Ancona), Italy

ABSTRACT

This paper presents a statistical technique to enhance vibration signals measured by laser Doppler vibrometry (LDV). The method has been optimised for LDV signals measured on bearings of universal electric motors and applied to quality control of washing machines. Inherent problems of LDV are addressed, particularly the speckle noise occurring when rough surfaces are measured. The presence of speckle noise is detected using a new scalar indicator kurtosis ratio (KR), specifically designed to quantify the amount of random impulses generated by this noise. The KR is a ratio of the standard kurtosis and a robust estimate of kurtosis, thus indicating the outliers in the data. Since it is inefficient to reject the signals affected by the speckle noise, an algorithm for selecting an undistorted portion of a signal is proposed. The algorithm operates in the time domain and is thus fast and simple. The algorithm includes band-pass filtering and segmentation of the signal, as well as thresholding of the KR computed for each filtered signal segment. Algorithm parameters are discussed in detail and instructions for optimisation are provided. Experimental results demonstrate that speckle noise is effectively avoided in severely distorted signals, thus improving the signal-to-noise ratio (SNR) significantly. Typical faults are finally detected using squared envelope analysis. It is also shown that the KR of the band-pass filtered signal is related to the spectral kurtosis (SK).

INTRODUCTION

Vibration measurements are widely used in industry for detection of mechanical defects. As the defects may arise during the manufacturing process or during the operating lifetime, vibration-based diagnostics is important for both quality control of commercial products and condition monitoring of industrial machines, such as motors, generators and pumps. Therefore, extensive research has been conducted to monitor the critical mechanical components, such as bearings [1] and [2].

Quality control is an important aspect of industrial production, particularly in the large market of household appliances. The aim of quality control is the distinction between good and faulty products by means of advanced diagnostic techniques. The most common vibration sensor used in fault diagnostics is the piezo-electric accelerometer. However, accelerometers appear to be less attractive for on-line quality control because of several limitations, such as invasiveness (local mass loading) and problematic installation on a tested product [3] and [4]. For this reason, the use of laser Doppler vibrometry (LDV) has become an increasingly popular technique [3], [4], [5], [6], [7], [8], [9], [10], [11], [12] and [13] owing to the non-contact principle of the laser.

Primarily, laser vibrometers offer flexible non-invasive measurements with high spatial resolution in a wide frequency range. In addition, LDV measurements can be easily automated, which significantly reduces the testing time and makes LDV particularly efficient for test stations in a production line environment. For example, laser vibrometers have been recently employed in detection of manufacturing defects in washing machines (WMs) [3] and electric motors [4] and [5].

Despite the advantages of LDV, vibration measurements on rough surfaces can be distorted by speckle noise [6], [7], [8], [9], [10], [11] and [12]. Hence this paper presents a novel technique for avoiding the speckle noise by selecting a portion of a signal which is undistorted. Although the discussion is focused on LDV signals from WMs and bearings of electric motors, the method has a potential to be used in various applications of laser vibrometry, as well as in diagnostics of other mechanical systems.

The paper is organised as follows. In Section 2, the principle of LDV is explained and inherent problems in measurement are discussed. In Section 3, a new statistical indicator kurtosis ratio (KR) is proposed and utilised for detection of speckle noise. Section 4 presents an algorithm for localisation of undistorted segments and provides instructions for optimising the algorithm parameters. The results of experimental validation are presented in Section 5. The conclusion is given in Section 6.

MEASUREMENT OF LDV SIGNALS

LDV is a non-contact measurement technique used to acquire the velocity (or displacement) of vibrating objects. Laser Doppler vibrometers and velocimeters (LDVs) typically use a helium-neon (HeNe) laser focused on the surface of a tested object. The surface reflects light from the laser beam and the Doppler frequency shift of the reflected light is demodulated to measure the vibration velocity, which is aligned with the axis of the incident laser beam [13].

Due to the non-contact principle of LDV, the measurement process does not affect the object of interest (mass loading is avoided), hence its natural stiffness and damping is unchanged [13]. LDVs are capable of measuring vibration in a wide frequency range: near DC to several MHz. The upper limit of the dynamic range depends on the maximum Doppler frequency which can be tracked; the lower limit is bounded by the smallest change in the Doppler frequency which can be detected by the instrument.

LDV operates reliably on clean and smooth surfaces where the amount of backscattered light is sufficient for subsequent signal processing. When the Doppler signal is insufficient, the demodulation unit in the vibrometer is unable to derive an accurate velocity waveform, resulting in signal distortion. The degree of severity in the signal distortion may vary from hardly noticeable deformations to a completely unintelligible signal.

The distortion may be caused by various sources, such as:

- *Imperfect focusing of lenses*.
- *Dust and debris*: When dust is present in the measurement environment, the laser beam undergoes multiple reflections and scattering, usually causing attenuation of the signal.
- *Distance from the object*: The performance of laser vibrometers is optimal when the stand-off distance is close to the coherence maxima of the light source.
- *Slanted surface*: When the laser beam hits a surface at a large incident angle, the beam is reflected sideways and the intensity of the Doppler signal is thus insufficient.

- *Speckle noise*: When measurements are conducted on a rough surface, undesired interference occurs and introduces uncertainty in the demodulation process [7] and [8]. This results in signal distortion which affects all types of laser vibrometers [7] and is referred to as speckle noise. This problem appears to be the most important limitation of LDV [8], [9] and [10].

SPECKLE NOISE

A *speckle pattern* is produced when coherent waves of an incident laser beam are dephased during backscatter from a surface that is rough on the scale of the optical wavelength [9] (Fig. 1(b)). Constructive and destructive interferences of these waves result in a random distribution of high and low intensities, respectively [9], which form a characteristic image shown in Fig. 2. As can be observed, the speckle pattern is a continuous array of bright and dark "spots" which have a grainy speckly appearance, hence the name speckle pattern [10].

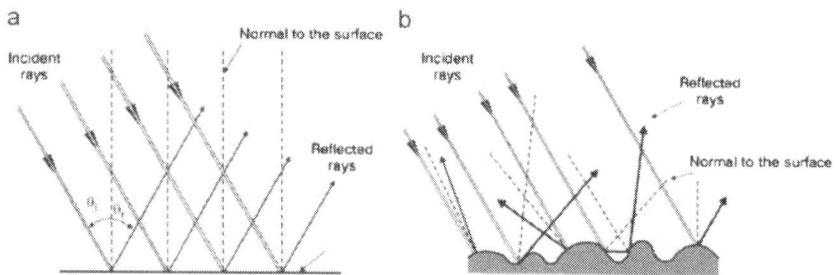

Figure 1. Dependency of LDV on optical properties of the measured surface: (a) specular reflection on a smooth surface, (b) diffuse reflection on a rough surface. These figures are copyright © by the Center for Occupational Research and Development (CORD). Used with permission of the copyright holder.

As the reflected waves have different phase, transitions between bright and dark speckles during the measurement generate noise in the output of the laser vibrometer. The amount of speckle noise depends on the rate of change in the speckle pattern and is thus considerable for surfaces with substantial in-plane motion component. Speckle noise has a pseudo-random character and is also referred to as pseudo-vibration [9], [10] and [11].

The presence of speckle noise in an LDV signal is characterised by large signal amplitudes (spikes) detectable by listening to the signal. Specifically, vibration signal can be converted to a sound waveform and played as an audio file, since the relevant diagnostic information is contained in the audible frequency range (20 Hz to 20 kHz). Speckle noise can then be perceived as a "scratching" sound, which resembles the impulsive noise of old gramophone recordings. However, this manual technique demands an experienced user and is inefficient for validating large signal databases.

Speckle noise can be reduced using a retro-reflective tape (or paint) in order to increase reflectivity of the measured surface [8]. Unfortunately, this technique appears to be impractical for automated quality control systems, as it involves additional consumption of material and time. The influence of surface roughness on the accuracy of LDV measurements was studied by Gasparetti and Revel [6]. Denman et al. [11] demonstrated that there exists an optimum position between the rotating target and the photodetector at which up to a 10 dB reduction in speckle noise can be achieved. Halkon and Rothberg [12] studied speckle noise characteristics in scanning LDV measurements on rotors.

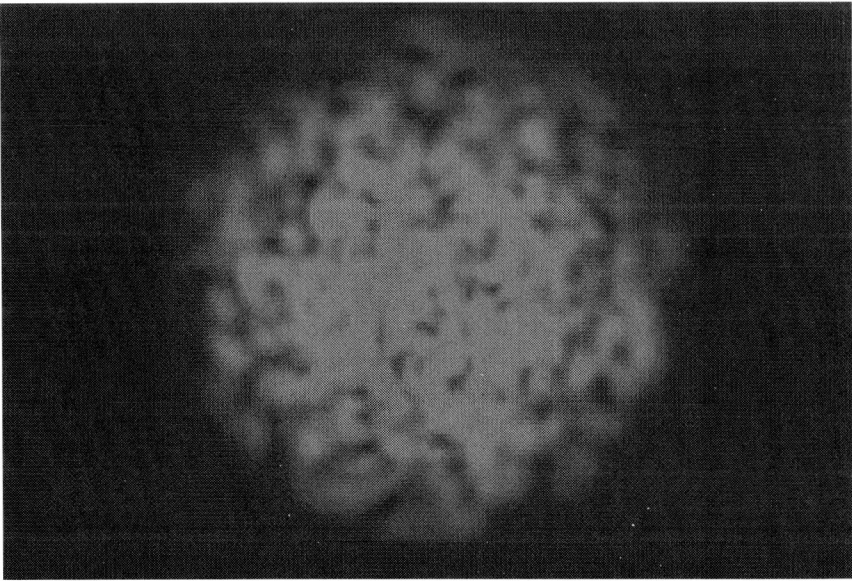

Figure 2. Example of the speckle pattern (courtesy of Vadim Makarov, Department of Electronics and Telecommunications, Norwegian University of Science and Technology (NTNU), Trondheim, Norway).

Measurements on electric motors

Fig. 3 presents a test station for quality control of electric motors. As can be seen, the shaft of the tested motor is vertically oriented in order to achieve constant load distribution and avoid modulations with the shaft speed. The nominal speed of the motor is 12 300 rpm (205 Hz). Each motor contains two bearings (SKF 6202), one in the lower part of the rotor and one in the upper part. The vibration velocity of each bearing is measured by an industrial single-point vibrometer [14]. It has been observed that speckle noise occurs more frequently in the upper bearing signal because the lower bearing is more stabilised by the weight of the motor.

Figure 3. Measurement setup for quality control of electric motors.

The measurement process consists of three steps: (1) accelerating the motor, (2) keeping it rotating at a constant speed for few seconds, (3) stopping the motor. The steady state of 2 s is then extracted and the resulting signal is stored in a database. Approximately 150 motors were measured, including 30 defective motors with a fault in the upper or the lower bearing. The signals were acquired by a 16-bit data acquisition board with a sampling frequency of 50 kHz. During the pre-processing, each steady-state signal is decimated by a factor of 2, hence the actual sampling frequency is 25 kHz.

Fault simulation

All tested motors are universal electric motors employed in automatic WMs. Universal motors are series-wound armature controlled motors capable of operating with either a DC or an AC power supply. The advantages of universal motors are inexpensiveness, high starting torque and ability to run at high speeds, which makes them the most common choice for household appliances (WMs, vacuum cleaners, hair dryers, food mixers), as well as power tools (drills, saws).

The goal is to reveal manufacturing defects in bearings caused by machines assembling the motors in a production line. Such machines may occasionally use an inappropriate force (under certain conditions) and cause a mechanical shock to a motor, thus damaging its bearings. The defects are induced by hammer shocks to the surface of the bearings with the aim to simulate the worst possible damage that can occur in reality. The fault simulation is performed by the manufacturer and a further methodology cannot be disclosed.

Table 1 gives parameters of the tested bearings, as well as the bearing defect frequencies (BDF)[1] and [2]. Since the contact angle φ is unknown, characteristic frequencies are estimated using the approximate formulae [2] which assume that $d/D \cdot \cos\varphi = 0.2$. This approximation is reasonable for most bearings with between 6 and 12 balls [2].

Table 1. Bearing parameters and expected fault frequencies

Bearing designation	SKF 6202
Bearing type	Deep groove ball bearing, single row
Number of balls	$N_b = 8$
Ball diameter	$d = 15\,mm$
Pitch circle diameter	$D = 35\,mm$
Contact angle	Unknown
Shaft rotation frequency	$f_r = 205\,Hz$ (12 300 rpm)
Fundamental train frequency	$FTF \approx 0.4 \cdot f_r = 82\,Hz$
Bass spin frequency	$BSF \approx 1.12 \cdot f_r = 229.6\,Hz$
Ball pass frequency on the outer race	$BPFO \approx 0.4 \cdot f_r \cdot N_b = 656\,Hz$
Ball pass frequency on the inner race	$BPFI \approx 0.6 \cdot f_r \cdot N_b = 984\,Hz$

Measurements on washing machines

The test station for quality control of WMs is displayed in Fig. 4(a). The WMs are of the front-loading type, hence the axis of the drum is in the horizontal position, as well as the axis of the motor (similar type as inSection 2.2). As shown in Fig. 4(b), the laser vibrometer is pointed to the drum of the WM and measures velocity of surface vibrations. The measurement is conducted during the spinning (centrifugal) phase of the wash cycle and consists of the run-up, steady state and run-down. Velocity signals are acquired for about 30 s and sampled at 20 kHz. WMs are lifted up during the measurement in order to guarantee insulation from the production line. For the same reason a seismic mass is used as a basement for the laser support. Each WM is tested with water inside the drum, which corresponds to a standard load of the pre-filling stage.

a   b

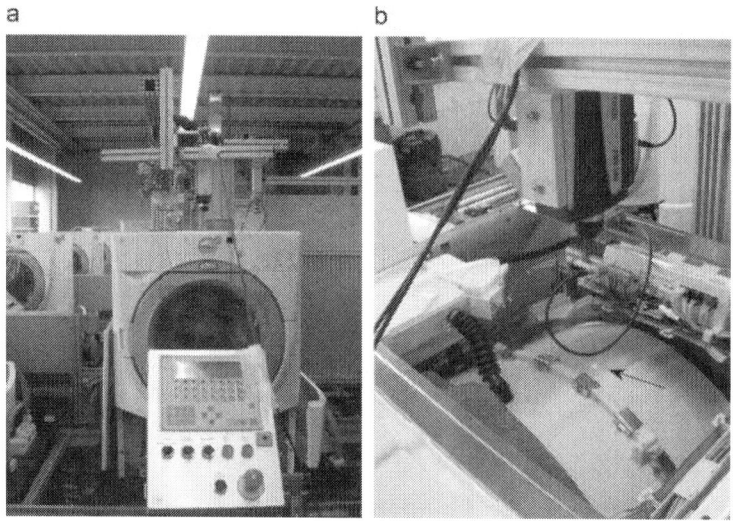

Figure 4. Measurement setup for quality control of washing machines.

General failures of WMs are caused by defects of its components, such as the pump, valves, screws, dampers, belt, counterweight and bearings (in our case SKF 6307 and 6206). However, the most frequent faults of WMs are related to the rotating parts of the motor, namely its bearings and rotor. The speed of the motor is again 205 Hz, hence the bearing fault frequencies are expected to be the same as in Table 1. The motor is connected to the drum by a belt drive, yielding a fixed ratio $f_r/f_0 = 13.5$, where f_r and f_0 are respectively the speed of the motor and the speed of the drum. Therefore, $f_0 = 205/13.5 \approx 15.2\,Hz$ (912 rpm).

Description of the method in Sections 3 and 4 is based on signals from electric motors (for which the method was developed first). Signals from WMs are analysed in Section 5.2.

DETECTION OF SPECKLE NOISE

This section presents a statistical technique for detecting the presence of speckle noise in LDV signals.Fig. 5 shows a measured signal distorted by a large amount of speckle noise. As can be observed, speckle noise has a character of randomly occurring impulses which increase the amplitude of the affected samples. This results in samples with abnormal amplitudes, i.e. *outliers*. In this paper, the term outlier refers to a signal sample which deviates significantly from the majority of samples in the signal. Specifically, the amplitude of such a sample does not follow the amplitude distribution of the signal and represents extreme deviation from the mean.

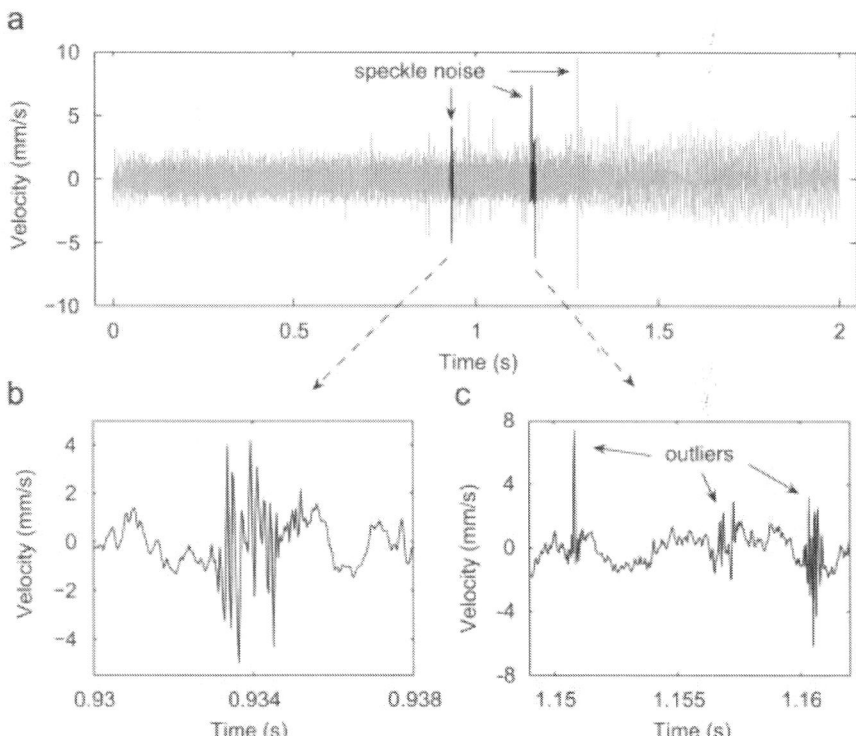

Figure 5. Speckle noise in an LDV signal (from the upper bearing of electric motor).

Since speckle noise is characterised by the presence of outliers, a specific indicator for detection of outliers will be proposed in Section 3.1. This indicator is based on kurtosis, as kurtosis is a measure of impulsiveness that can effectively quantify the impulsive character of speckle noise. Kurtosis is defined as the fourth central statistical moment μ_4 normalised by the square of the second central moment μ_2 (i.e. the variance) [15], [16], [17] and [18]:

$$\beta_2(x) = \frac{\mu_4}{(\mu_2)^2} = \frac{E\{(x - \mu_x)^4\}}{\sigma_x^4},$$

(1)

Where μ_x and σ_x is the mean value and the standard deviation of x, respectively, and $E\{f(x)\}$ denotes the expectation value of $f(x)$. Note that the power of four in the numerator gives considerable weight to high amplitudes.

By definition, kurtosis is 3 for the normal (Gaussian) distribution[1] and increases with increasing proportion of high amplitudes in the data. This property is commonly used in bearing diagnostics [17], [18], [19],[20] and [21], as peak amplitudes in vibration signals correspond to periodic impulses caused by mechanical defects. Fig. 6 shows the probability density function (PDF) corresponding to a simulated vibration signal from a good (a) and faulty (b) bearing. In the case of the good bearing, the simulated signal is stationary and follows the Gaussian distribution. On the other hand, the damaged bearing is characterised by a non-Gaussian distribution with dominant tails, as illustrated by the arrows in Fig. 6(c).

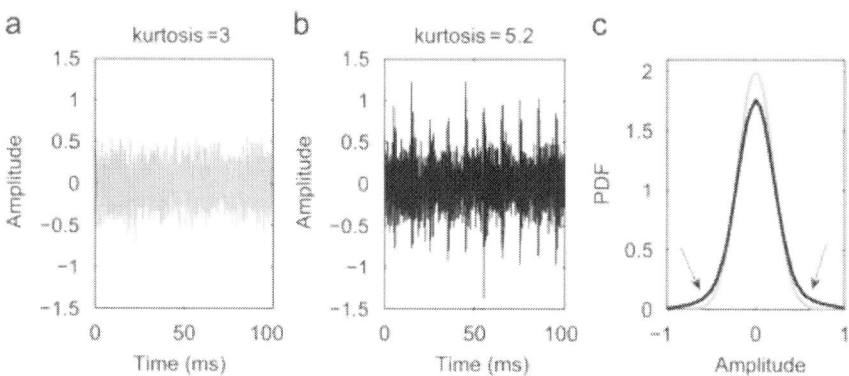

Figure 6. Ability of kurtosis to indicate the presence of impulses: (a) signal without impulses, (b) impulsive signal from a faulty bearing, (c) corresponding estimates of the PDF.

In our method, kurtosis is employed only for localisation of *random* impulses caused by the speckle noise (Fig. 5) and not for detection of *periodic* impulses related to bearing faults (Fig. 6(b)). Nevertheless, the influence of the fault-related impulses is examined in 3.1 and 4.5.

Kurtosis ratio

Kurtosis ratio is a new statistical indicator specifically designed for detection of speckle noise in LDV signals. The KR is a scalar expressing the amount of outliers in the data and is defined as

$$KR(x) = \frac{\beta_2(x)}{\beta_2(x_t)},$$

(2)

Where x is the raw (measured) signal, x_t is the *trimmed signal* and $KR(x)$ is the KR of x. The computational process of $KR(x)$ is illustrated in Fig. 7. The nomenclature is given in Table 2, where ICDF is an abbreviation of inverse cumulative distribution function.

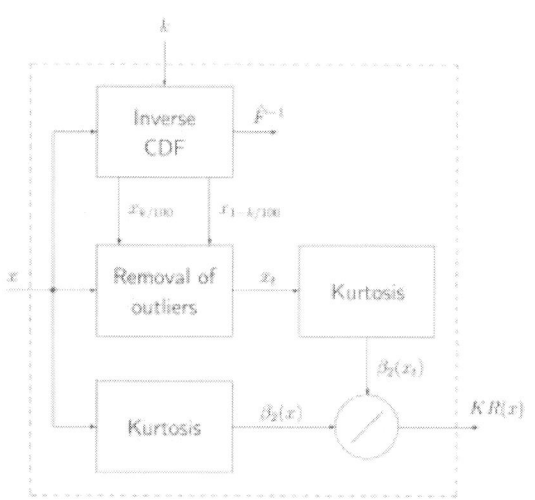

Figure 7. Computation of the kurtosis ratio (KR).

Table 2. Nomenclature for the computation of the KR

Symbol	Description
x	Raw signal
x_t	Trimmed signal
k	Outlier removal level (%)
$\hat{F}^{-1}$	Estimate of the ICDF
$x_{k/100}$	kth percentile of x
$x_{1-k/100}$	$(100-k)$th percentile of x
$\beta_2(x)$	Kurtosis of the raw signal
$\beta_2(x_t)$	Kurtosis of the trimmed signal
$KR(x)$	Kurtosis ratio of the raw signal

The trimmed signal x_t is obtained from x by removing the outlying samples caused by speckle noise (details of this procedure are given in Section 3.2). As a result, the "trimmed kurtosis" $\beta_2(x_t)$ is unaffected by the presence of outliers and thus represents a robust estimate of kurtosis. On the other hand, "standard kurtosis" $\beta_2(x)$ can be biased by outliers in x and increases significantly with increasing amplitudes of random impulses. Note that $\beta_2(x_t)$ was inspired by the trimmed mean in [16] and can be viewed as an extension of the same idea to the fourth central moment.

When no speckle noise is present in x, values of $\beta_2(x)$ and $\beta_2(x_t)$ are comparable and thus $KR(x) \approx 1$. On the contrary, $KR(x)$ for a distorted signal is much greater than one, because outliers increase only $\beta_2(x)$, while the robust estimate $\beta_2(x_t)$ is unaffected (unbiased). Speckle noise is detected when $KR(x)$ exceeds the threshold $thr_{KR} = 2$.

The threshold thr_{KR} controls the sensitivity of the detector to the speckle noise. The recommended value $thr_{KR} = 2$ indicates that outlying samples have biased $\beta_2(x)$ by a factor of 2. Provided that the trimmed signal is approximately Gaussian ($\beta_2(x_t) \approx 3$), the value $KR(x) = 2$ can be interpreted as the transition of the raw signal from the Gaussian distribution to the Laplace distribution, as their theoretical values of kurtosis are 3 and 6, respectively [15]. Indeed, the Laplace distribution belongs to supergaussian (leptokurtic) distributions [22] that are assumed for data with impulsive components [23], as suggested by the PDF curve in Fig. 6(c).

The definition of the KR guarantees that signals from good and faulty bearings are treated equally, i.e. ability of $KR(x)$ to detect the speckle noise is insensitive to the presence of the fault-related impulses. This is achieved owing to the primary assumption that the impulses to be detected are random and hence corresponding to outlying samples. Since the fault-related impulses are periodic, their proportional representation is too high to be completely removed as outliers. As a consequence, periodic impulses are present in the trimmed signal x_t, resulting in an increased value of $\beta_2(x_t)$. Since $\beta_2(x)$ is also increased, KR remains below the threshold $KR(x) < thr_{KR}$ and the signal is thus regarded as correctly measured despite the presence of the fault-related impulses. The KR of the signal in Fig. 6(b) is $KR(x) = \beta_2(x)/\beta_2(x_t) = 5.2/3.9 \approx 1.33$.

Removal of outliers
The removal of outlying samples is achieved by thresholding the raw signal x:

$$xt = (thr_{low} < x[n] < thr_{high}), \tag{3}$$

$$thr_{low} = x_{k/100} = x_{0.05} = \hat{F}^{-1}(0.05), \tag{4}$$

$$thr_{high} = x_{1-k/100} = x_{0.95} = \hat{F}^{-1}(0.95), \tag{5}$$

where thr_{low} and thr_{high} is the lower and the upper threshold, respectively. In other words, the trimmed signal xt consists only of samples $x[n]$ satisfying Eq. (3). The remaining samples are simply excluded from the signal since the resulting non-stationarities in xt do not affect the value of $\beta_2(xt)$ (kurtosis describes only the distribution of amplitudes in a signal).

Outlier removal is performed automatically since both thresholds are computed directly from the samples of the given signal. The numeric value of each threshold depends on the chosen percentage point k, where $k = 1, 2, \ldots, 99\%$. This parameter has been termed *outlier removal level*, since it is an integer specifying the amount of samples to be removed. For $k = 5\%$ (recommended value), the trimmed signal xt is obtained from x by excluding 5% of the lowest samples and 5% of the highest samples. This is achieved by selecting thr_{low} and thr_{high} as the 5th and the 95th percentile of x, respectively. The values of both percentiles are returned by the ICDF for a given x and k, as shown in Fig. 7.

Fig. 8(a) displays the signal from Fig. 5(a) and the computed thresholds. Fig. 8(b) depicts an estimate $\hat{F}^{-1}(x)$ of the ICDF using one realisation of the signal. This function is obtained by sorting the samples of x in ascending order, and then scaling the time axis by the total number of samples in order to obtain values of probability from zero to one. Probabilities 0.05 and 0.95 correspond to the 5th percentile $x_{0.05}$ and the 95th percentile $x_{0.95}$, respectively, which determine the lower and the upper threshold, respectively. These thresholds guarantee that a *fixed percentage* of low and high amplitudes is removed, depending on the chosen k. For this reason, the character of impulsive noise can be arbitrary because the computation of thresholds does not depend on the amplitude of the impulses or their total number. As a result, speckle noise is reliably detected in a large variety of LDV signals.

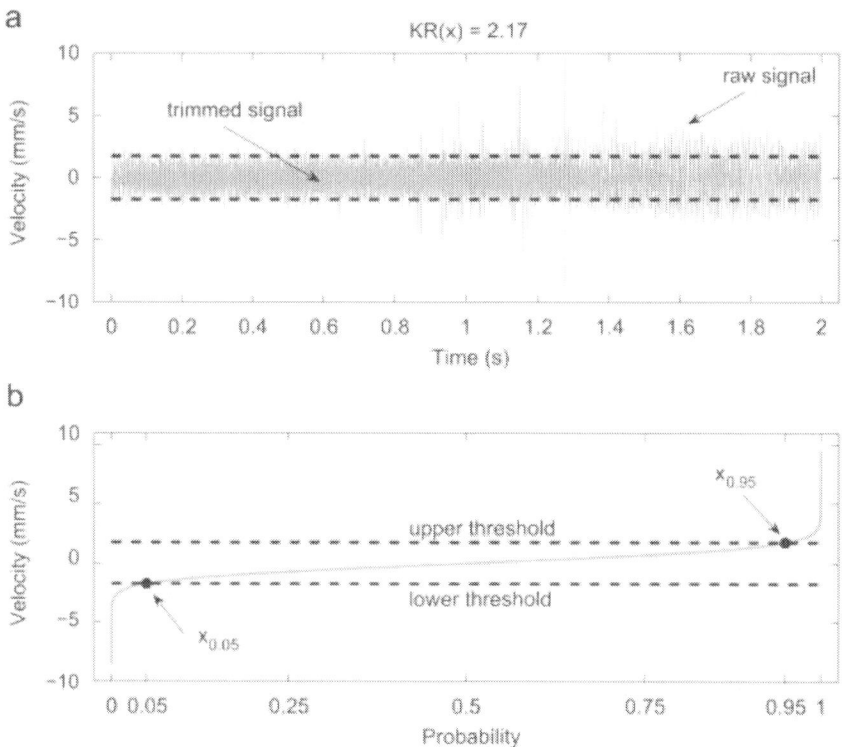

Figure 8. Automatic removal of outlying samples: (a) thresholding of the raw signal, (b) computation of thresholds using the ICDF.

The removal of outliers is efficient when both thresholds thr_{low} and thr_{high} are located beyond the inflection points of the ICDF (Fig. 8(b)). For a smaller k (e.g. $k=2\%$), some outliers may still be present in the trimmed signal, thereby contributing to a bias of $\beta_2(x_t)$. As a consequence, $KR(x)$ would be decreased and some undesired impulses may thus be undetected. The selection $k=5\%$ should be adequate for most applications since 10% of removed samples will certainly comprise the majority of potential outliers in the data. Our signal database was statistically evaluated in order to confirm that the chosen value of k is sufficient for typical shapes of the ICDF.

SELECTION OF AN UNDISTORTED REGION

Strict time constraints are imposed on quality control processes in production lines, and since obtaining multiple measurements is very time-consuming, signals affected by speckle noise cannot be simply rejected and measured again. Therefore, distorted signals must be further processed in order to select a portion of a signal without any undesired components.

Description of the algorithm

An algorithm for avoidance of speckle noise is shown in Fig. 9. The nomenclature is given in Table 3. The input of the algorithm is a raw LDV signal x; the output is the selected region r. When no speckle noise is detected, then $r=x$. The algorithm operates solely in the time domain and is detailed in the following subsections.

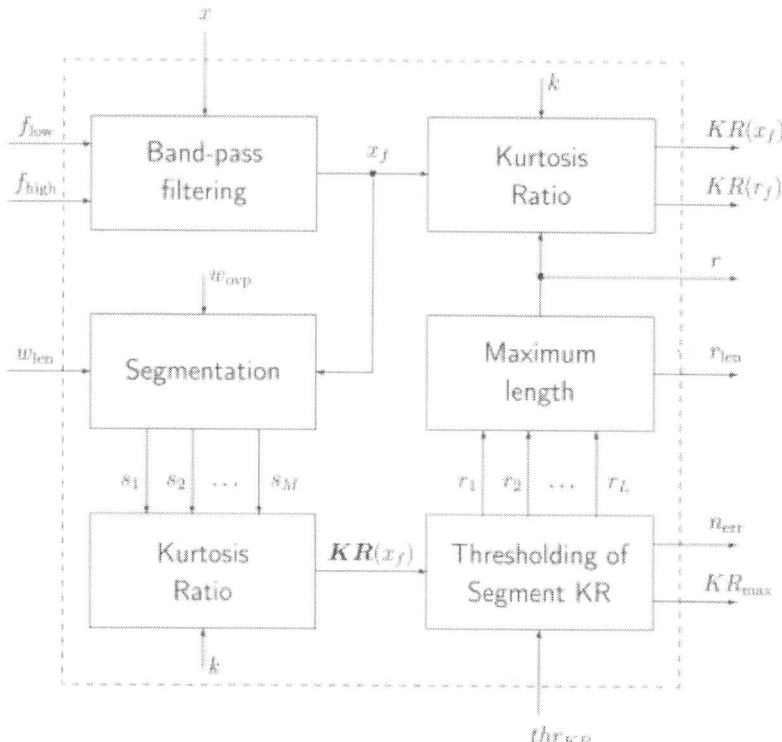

Figure 9. Block diagram of the algorithm for avoidance of speckle noise.

Table 3. Nomenclature for the algorithm in Fig. 9

Symbol	Description
x	Raw (measured) signal
xf	Filtered signal
sm	mth segment of $_{xf}$, m=1,2,...,M
rl	lth candidate region, l=1,2,...,L
r	Selected region (a portion of x)
flow	Lower frequency edge (Hz)
fhigh	Upper frequency edge (Hz)
wlen	Window length (samples)
wovp	Window overlap (80%)
k	Outlier removal level (5%)
thr_{KR}	Threshold for the KR (2 by default)
$KR(r_f)$	Kurtosis ratio of the filtered region
$KR(x_f)$	Kurtosis ratio of the filtered signal
$\mathbf{KR}(_{xf})$	Segment kurtosis ratio of the filtered signal
KRmax	Maximum value of $\mathbf{KR}(_{xf})$
rlen	Length of the selected region (%)
nerr	Number of detected errors

Band-pass filtering

A digital band-pass filter is designed according to the desired band $_{fBW}= \langle_{flow,fhigh}\rangle$, where $_{flow}$ and $_{fhigh}$ is the lower and the upper frequency edge, respectively. The band from 5 to 10 kHz has been selected as discussed in Section 4.2.

Segmentation

The filtered signal $_{xf}$ is divided into overlapping segments using a rectangular window. Theoretically, the window length $_{wlen}$ should be maximised since the confidence of statistical indicators increases with an increasing number of samples. On the other hand, segments should not be excessively long as good time resolution is necessary to detect impulses of short duration. Therefore, high overlapping of segments is used ($_{wovp}=80\%$) in order to achieve both sufficient window length (for reliable estimation of kurtosis) and accurate time resolution. Further details on the segmentation are given in Section 4.5.

Segment Kurtosis ratio

The Kurtosis ratio is computed for each segment of the filtered signal:

$$KR_m = KR(s_m), \quad m = 1, 2, \ldots, M, \tag{6}$$

$$\boldsymbol{KR}(xf) = [KR1, KR2, \ldots, KRM], \tag{7}$$

where sm is the mth segment of xf, M is the total number of segments and $\boldsymbol{KR}(xf)$ is the segment KR of xf. It is important to distinguish between $\boldsymbol{KR}(xf)$ and $KR(xf)$. The former is a vector whose mth element is the KR computed in the m th segment of xf, whereas the latter refers to a single value of the KR computed from all samples of xf, i.e. $KR(xf) = \beta_2(xf)/\beta_2(xtf)$, where xtf is the trimmed version of the filtered signal.

$\boldsymbol{KR}(xf)$ is very sensitive to spectral changes caused by the speckle noise, as can be seen in Fig. 10. Computation of a spectrogram by the short-time Fourier transform (STFT) is not included in the algorithm, but is always recommended for visualisation purposes. $\boldsymbol{KR}(xf)$ in Fig. 10(c) reaches very high values and is therefore "zoomed" to the vicinity of the threshold $thr_{KR} = 2$.

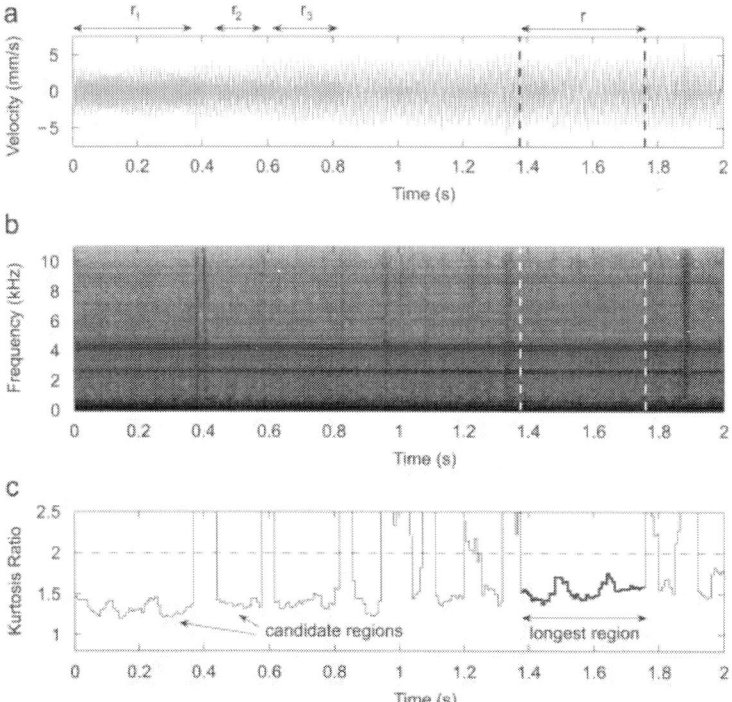

Figure 10. Example 1—sensitivity of the kurtosis ratio to spectral changes caused by the speckle noise: (a) raw signal x, (b) spectrogram of x , (c) thresholding of $\boldsymbol{KR}(xf)$ to obtain the longest undistorted region.

Thresholding of segment KR

$KR(_{xf})$ is thresholded to obtain a set of candidate regions, which consist of the signal segments fulfilling the condition $KR(_{sm}) < thr_{KR}$. The region comprising the highest number of segments (i.e. the longest region) is selected as the best candidate for an undistorted portion of the signal. The resulting region is depicted in Fig. 10(c), but remember that r is a portion of x.

Indicators of signal distortion

Three parameters are computed in order to indicate the degree to which the raw vibration signal is distorted.

- $KR(_{xf})$—kurtosis ratio of the filtered signal. As will be shown in Section 4.3, $KR(_{xf})$ is a more sensitive indicator than $KR(x)$ and is hence preferred to quantify the overall distortion of the raw signal.
- KR_{max}—the maximum value of the segment KR, i.e. $KR_{max} = max\{KR(_{xf})\}$. This indicator quantifies the largest error and is useful for localising the most distorted segments.
- $nerr$—number of measurement errors. Each error is detected when $KR(_{xf})$ exceeds the threshold thr_{KR}.

Indicators of the selection result

Finally, the system computes two indicators expressing the quality of the selected region.

- $KR(_{rf})$—kurtosis ratio of the filtered region $_{rf}$ (i.e. a portion of $_{xf}$ corresponding to the selected region). This indicator is more sensitive than $KR(r)$, hence its small value provides a better assurance that the region r contains no speckle noise. When $KR(_{rf}) > 2$, then r is regarded as distorted and it is either impossible to avoid speckle noise or the parameters of the algorithm are incorrect.
- $rlen$—length of the selected region expressed as a percentage of the raw signal length. Selection is considered successful when $rlen \geqslant 25\%$ for signals from electric motors (Section 2.2) or $rlen \geqslant 10\%$ for signals from washing machines (Section 2.3).

Design of the band-pass filter

This section presents an analysis procedure that was performed by the authors to design the band-pass filter used in Section 4.1.1. Specifically, a suitable frequency band for filtering was determined by spectral analysis of resonance frequencies. Although demonstrated using a single LDV signal, similar characteristics were observed in the majority of signals in our database (Section 2.2). It is suggested that a user should perform similar analysis in order to select an optimal band for a particular class of

LDV signals. Alternatively, the band can be determined using the spectrogram as shown in Section 5.2.

Fig. 11(a) presents amplitude spectra of an LDV signal measured on a good motor and unaffected by the speckle noise. The spectra are computed by discrete Fourier transform (DFT) and autoregressive (AR) modelling [24]; in both cases the depicted spectrum is an average of spectra computed in 20 non-overlapping segments of the signal. The segment length is 256 and the AR model order p is 20. As can be noticed, both the DFT and AR spectra exhibit a large peak in the low-frequency range (around 500 Hz). This peak represents the resonance frequency of the motor and is thus unimportant for diagnostics of bearing faults. For this reason, it is advantageous to attenuate the band up to 1 kHz and estimate the spectrum of the high-pass filtered signal, as shown in Fig. 11(b). Note that a similar technique was used by Martin and Honarvar [18].

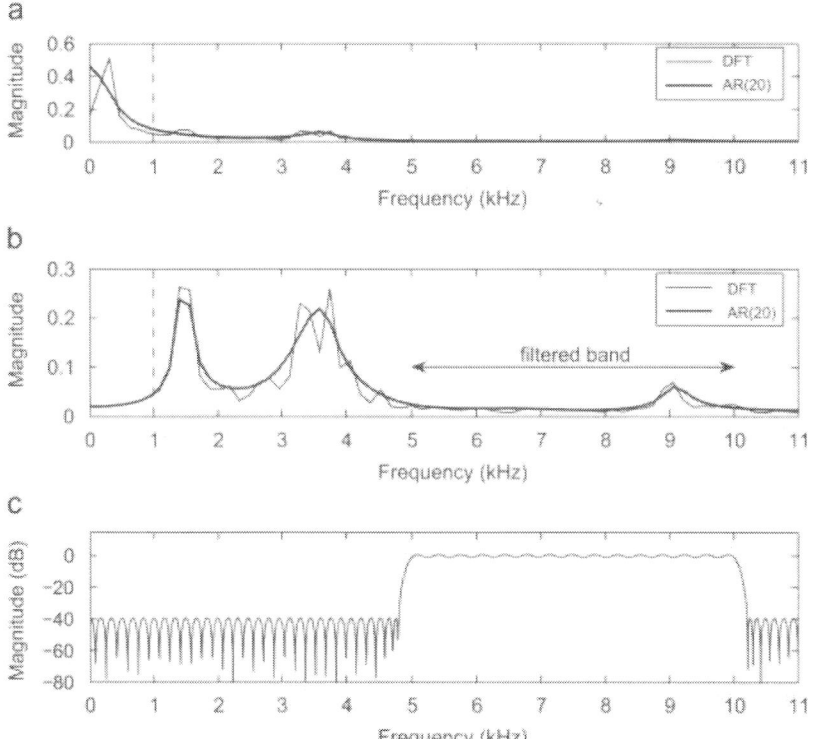

Figure 11. Selection of the filtered band: (a) spectrum of the full-band signal, (b) spectrum of the high-pass filtered signal, (c) resulting band-pass filter.

When the dominant resonance is removed, the minor peaks located at 1.5, 3.5 and 9 kHz are considerably more visible. These peaks correspond to vibration modes of other structural components, including the bearing housing. Similar resonance frequencies were found by analysing signals from good and faulty motors, hence it is assumed that these resonances are independent of bearing faults and are probably excited by stationary background noise.

The band 5–10 kHz was intentionally chosen for detection of speckle noise for two reasons. First, as revealed by the spectrogram in Fig. 10(b), each LDV error is attended by a short random impulse whose energy is distributed over the whole spectrum. Nevertheless, the energy of the impulse is more visible in the upper frequency range, which makes the band 5–10 kHz particularly sensitive to the impulses caused by speckle noise. Second, several authors [17], [18], [19], [20] and [21] have concluded that statistical indicators (such as kurtosis and crest factor) are generally more efficient when calculated in frequency bands since reliability of global (full-band) indicators tends to be inhibited by various masking effects[19] and [21].

The filter is designed using the Remez exchange algorithm [25] in order to obtain a linear-phase FIR (finite impulse response) filter with an equiripple frequency response (Fig. 11(c)). The following specifications are used: transition bandwidth 200 Hz, passband ripple $\delta_p = 0.1$ (1.74 dB), stopband ripple $\delta_s = 0.01$ (40 dB attenuation). The filter order to meet the specifications is around 300.

The effect of band-pass filtering

The advantages of band-pass filtering are illustrated in Fig. 12. The raw signal x contains no distinguishable impulses and its ICDF has a typical shape for signals without outliers. On the other hand, random impulses are clear in the signal x_f filtered in the range 5–10 kHz. As a consequence, the presence of outlying samples modifies the shape of the ICDF, resulting in a steep slope at the edges corresponding to low and high amplitudes.

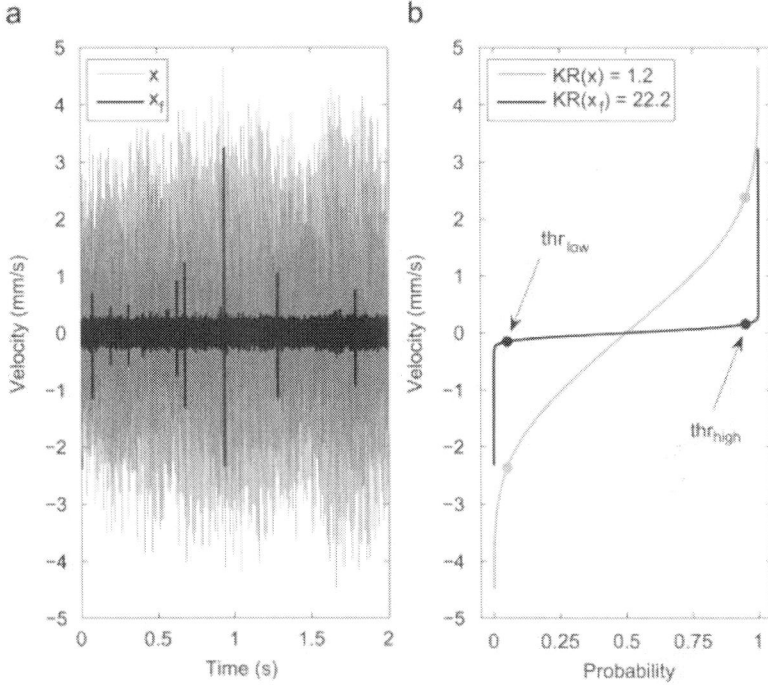

Figure 12. Improvement in the sensitivity of the KR by band-pass filtering: (a) raw and filtered signal, (b) corresponding estimates of the ICDF.

The band-pass filter plays a fundamental role in the algorithm, as can be demonstrated by comparing the KR of x and $_{xf}$. While the former exhibits a small value of $KR(x)=1.2$, the latter has reached an excessive value of $KR(_{xf})=22.2$. Therefore, speckle noise would be detected only in the latter case, in which the KR exceeds the threshold $thr_{KR}=2$. This implies that filtering is essential to the ability of the KR to effectively detect the random impulses masked by resonance frequencies.

Comparison with spectral kurtosis

It is interesting to compare the KR with the spectral kurtosis (SK). The SK was introduced by Dwyer [26] and has been recently generalised by Antoni and Randall to a concept of kurtogram [21]. As the name implies, the SK is a function of frequency with each value corresponding to the kurtosis in a certain frequency band. On the other hand, $KR(_{xf})$ is a ratio of $_{\beta 2}(_{xf})$ and $_{\beta 2}(_{xtf})$, where $_{\beta 2}(_{xf})$ is the kurtosis of the band-pass filtered signal. Therefore, $_{\beta 2}(_{xf})$ is a measure of the impulsiveness of a signal in a limited frequency range and hence carries similar information as the SK in this range. The difference is that $_{\beta 2}(_{xf})$ is obtained using a wide band-pass filter that has been optimised in advance by spectral analysis (Fig. 11), whereas the SK can be used as an analysis tool to select the optimal band-pass filter within a bank of narrow-band filters [27]. Nevertheless, both techniques are intended to detect hidden non-stationarities in a signal.

Optimal window length

Instructions for choosing a proper window length are based on a repetition rate of bearing fault impulses. As mentioned earlier, a key issue concerning the KR is the ability to distinguish the bearing impulses from the impulses caused by speckle noise. Since the bearing impulses are periodic, it is necessary to guarantee that their proportional representation within each segment is higher than the amount of expected outliers (i.e. 10% for k=5%). Therefore, the window length depends on periodicity of the bearing impulses and their decay rate (damping). It is suggested that the window length should be at least four times greater than the highest characteristic period $1/f_{BDF}$

$$W_{len} \geq \frac{4 \cdot f_s}{f_{BDF}},$$
(8)

where f_s is the sampling frequency and f_{BDF} is the smallest of the expected fault frequencies, i.e. the cage frequency FTF (Table 1). For instance, if $f_s = 25\,kHz$ and $f_{BDF} = 80\,Hz$, the characteristic period $1/f_{BDF} = 12.5\,ms$ requires a minimal window length of 50 ms (1250 samples). Alternatively, this requirement can be expressed as a minimum number of complete shaft rotations per segment, which is 10 in this case $(50\,ms \cdot 205\,Hz = 10.25)$. The time resolution of the method depends on the window step $_{wstep} = _{wlen} \cdot (1 - _{wovp}/100)$, which is 250 samples (10 ms). The window length$_{wlen} = 1250$ will be used for signals in Section 5.1.

Simulated bearing signal

The numerical example above is depicted in Fig. 13, which presents validation of a simulated vibration signal. The signal is generated as a quasi-periodic train of exponentially decaying impulses with the cage frequency FTF = 80 Hz, three resonances (2, 6 and 8.5 kHz) and 2% random fluctuations in the spacing of the impulses. It should be noted that although the FTF usually appears as a modulating frequency, it is purposely used here to represent a worst case scenario, since the higher ball-pass frequencies (BPFO and BPFI) would tend to have lower kurtosis because of their closer spacing. Also note that the quasi-periodic model [28] was originally proposed to incorporate slipping of rolling elements, which is of smaller intensity in our bearing signals due to the purely axial load. Nevertheless, randomness is introduced by circulation of the defect around the measurement point, thus justifying the use of this model. Finally, the signal is distorted by additive white Gaussian noise for an SNR of $-6\,dB$ (Fig. 13(a)).

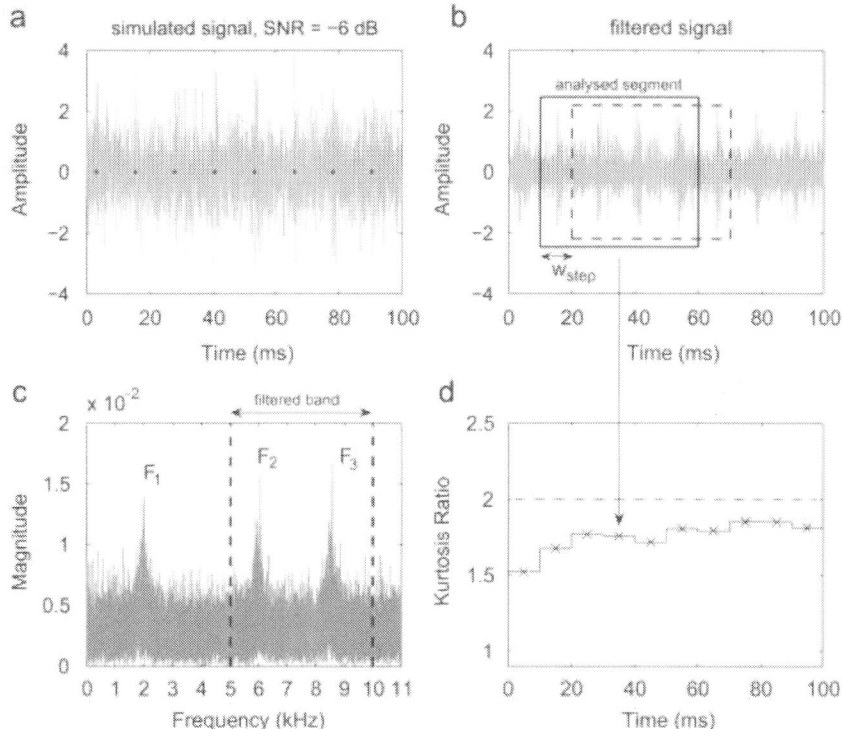

Figure 13. Selection of the optimal window length: (a) noisy impulse signal x , (b) segmentation of $_{xf}$, (c) amplitude spectrum of x , (d) segment kurtosis ratio of $_{xf}$.

Fig. 13(c) shows the amplitude spectrum of the noisy signal with smeared resonances due to imperfect periodicity of the impulses. Fig. 13(b) depicts the filtered signal with enhanced impulses and illustrates the segmentation process. As the impulses comprise more than 10% of each signal segment, $KR(_{xf})$ in Fig. 13(d) remains below the threshold for detection of speckle noise. Therefore, no segments are regarded as distorted and the complete signal x=r is preserved for detection of bearing faults.

Adjustment of the threshold
The example in Fig. 13 represents a maximum (theoretical) SNR up to which $KR(_{xf})$ remains always below the threshold $thr_{KR} = 2$ (the experiment was repeated for 100 realisations of the noise). For lower SNRs, bearing impulses gradually disappear in the noise, hence the amplitude distribution becomes Gaussian and $KR(_{xf})$ fluctuates between 1.2 and 1.6. For larger SNRs, the impulses dominate the signal and increase $KR(_{xf})$ up to 3.5 (average value for the clean impulse train). For

this reason, bearing impulses must be sufficiently buried in the noise, otherwise their presence may increase $KR(x_f)$ above the threshold and cause a failure of the method in the sense that signal segments are rejected due to a bearing fault.

Therefore, in applications where large bearing impulses are expected, thr_{KR} should be increased accordingly, e.g. to 3. As will be shown for actual LDV signals, $KR(x_f)$ is often greater than 4, hence this threshold would still detect the most severe speckle noise. On the other hand, some minor random impulses might be allowed into the selected region, but their influence seems negligible for correct fault diagnosis. The threshold of 2 appears optimal for our data in which the bearing impulses are masked by strong noise, thus requiring an extraction using the envelope technique (Section 5.1.4) or denoising based on the Morlet wavelet [5], [23] and [29].

EXPERIMENTAL RESULTS

Table 4 presents the results for six LDV signals measured on universal electric motors and WMs. Examples 1–3 are intended to illustrate varying severity of speckle noise, hence the corresponding signals were measured on good bearings (no fault-related impulses are expected). Examples 4–6 represent faulty cases and their purpose is to show that speckle noise is successfully avoided despite the presence of mechanical faults.

Table 4. Experimental results (EM, electric motor; WM, washing machine)

Example	1	2	3	4	5	6
Tested product	EM	EM	EM	EM	WM	WM
Condition	Good	Good	Good	Faulty	Faulty	Faulty
Fault location	–	–	–	Bearing	Motor	Motor
Measured bearing	Upper	Upper	Upper	Lower	–	–
KR_{max}	47.1	46.7	82.2	6.7	19.4	12.6
$KR(x_f)$	82.4	37.9	44.5	2.1	27.2	8.7
$KR(r_f)$	1.57	1.41	1.33	1.58	1.11	1.32
n_{err} (number of errors)	9	5	21	9	71	61
r_{len} (region length)	19.20%	31.60%	4.40%	32.20%	14.70%	12.40%

The majority of signals from electric motors contain only a small number of random impulses, hence the examples presented here are exceptional and represent the cases for which it is necessary to find the undistorted region. On the other hand, all measurements conducted on the drum of washing machines are affected by very severe speckle noise that appears to be inevitable. This is caused by two factors. First, the rotating drum vibrates in various directions, particularly during the run-up and run-down. Second, surface characteristics [6] of the drum are worse than those of the bearing housing, namely the reflective power is lower and the optical roughness is larger. This can be deduced from Table 4 showing that signals from WMs contain more measurement errors and undistorted regions are shorter.

Signals from electric motors

Example 1—acceptable case

An introductory example was shown in Fig. 10. Although speckle noise is avoided ($KR_{(rf)}=1.57$), the length of the selected region is insufficient (19.2%), but not critical. Severe distortion of the raw signal is clearly indicated by extreme values of $KR_{max}=47.1$ and $KR_{(xf)}=82.4$.

Example 2—successful case

The second example in Fig. 14 represents a successful case, in which the selected region is undistorted ($KR_{(rf)}=1.41$) and the region length is satisfactory (31.6%). As can be observed, band-pass filtering highlights the impulsive components, although the raw signal also exhibits the random impulses. A small number of errors ($n_{err}=5$) is a consequence of $KR_{(xf)}$ persisting above the threshold from 0.9 to 1.8 s (Fig. 14(d)). This signal was introduced in Figs. 5 and 8(a).

Example 3—short region

Filtering is necessary in this example since measurement errors are considerably less evident. Impulses can be clearly seen only in Fig. 15(c), since the wide-band spectrum is dominated by structural resonances. Localising an undistorted region is difficult due to many errors ($n_{err}=21$). Only a short region is available (4.4%), resulting in a need to slightly shift the position of the laser beam and measure the signal again [8]. Otherwise the human operator should be informed about a small amount of correct data. Nevertheless, the selected region is undistorted ($KR_{(rf)}=1.33$) as no scratching was heard during the listening test (Section 2.1).

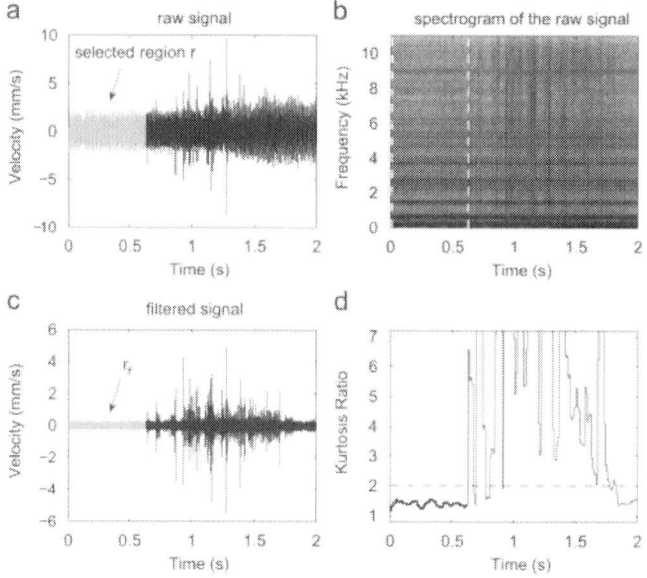

Figure 14. Example 2—selection of a sufficient region: (a) raw signal x, (b) spectrogram of x, (c) filtered signal $_{xf}$, (d) thresholding of $KR(_{xf})$.

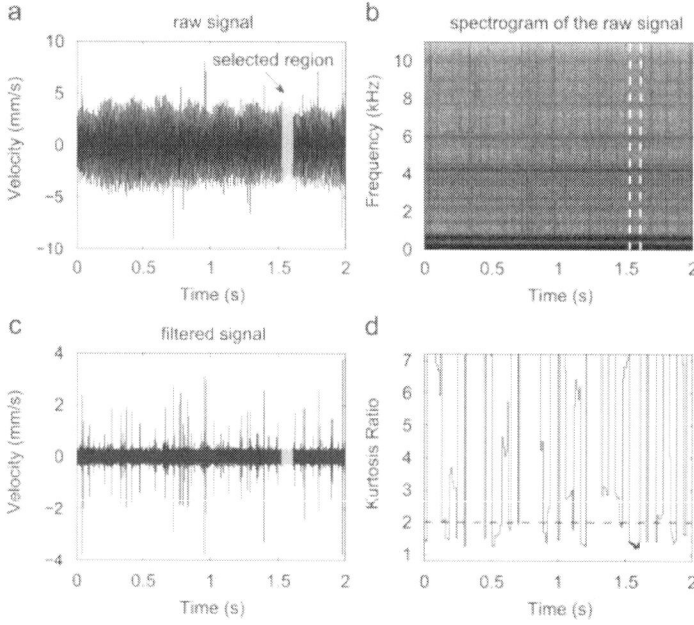

Figure 15. Example 3—selection of a short region: (a) raw signal x, (b) spectrogram of x, (c) filtered signal $_{xf}$, (d) thresholding of $KR(_{xf})$.

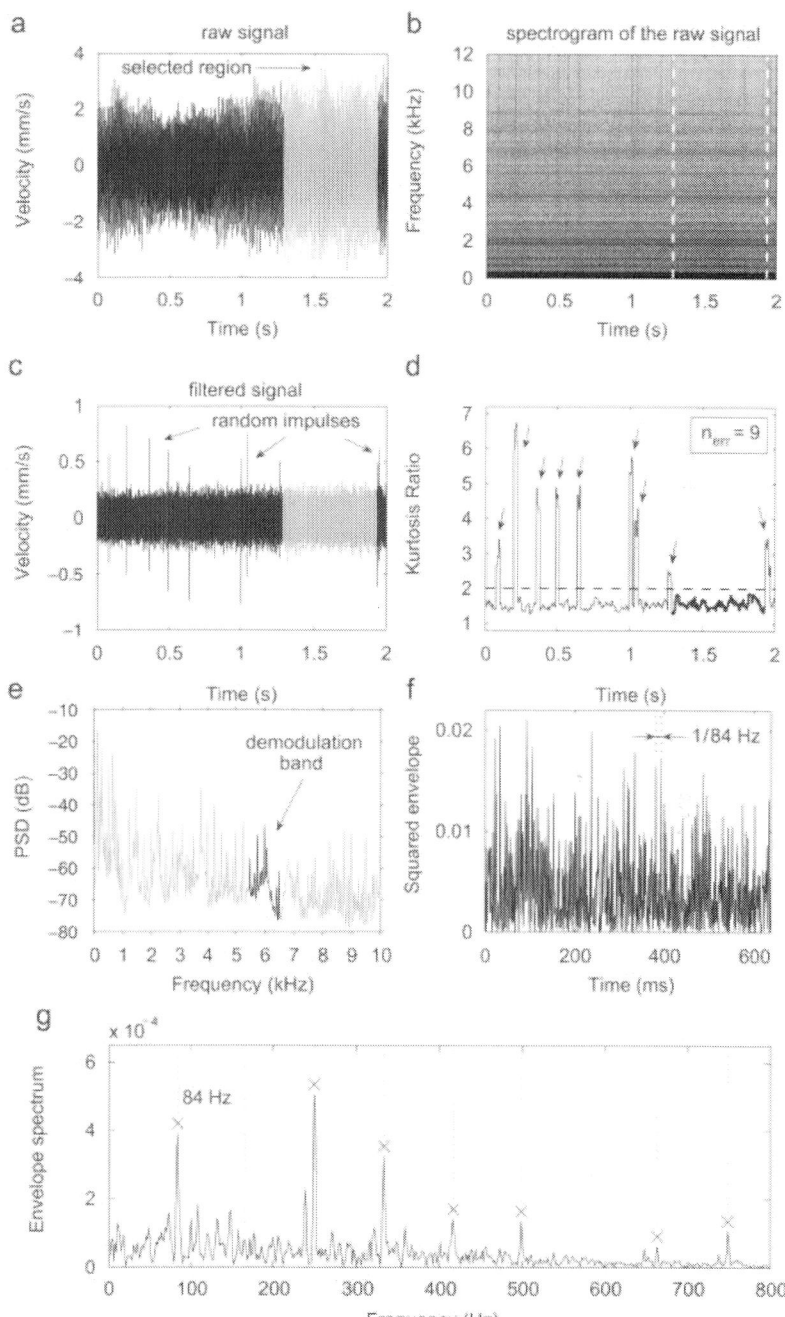

Figure. 16. Example 4—analysis of a faulty bearing: (a) raw signal x, (b) spectrogram of x, (c) filtered signal $_{xf}$, (d) $\mathbf{KR}(_{xf})$, (e) PSD of r, (f) squared envelope obtained from r, (g) magnitude of the squared envelope spectrum.

Example 4—faulty bearing

As mentioned in Section 2.2, speckle noise is generally less severe and less frequent in our signals from the lower bearings. Indeed, $KR(_{xf})=2.1$ is only slightly above the default threshold, thus the noise would be missed by the basic detector of Section 3 when using $thr_{KR}=3$. The filtered signal in Fig. 16(c) contains several random impulses partially visible in the spectrogram. Each impulse is associated with a sudden increase of $\boldsymbol{KR}(_{xf})$, thus $_{nerr}=9$ quantifies the number of random impulses in $_{xf}$. It is seen that these impulses are indeed unusual and random, which is in agreement with speckle noise being a pseudo-vibration component with no time constant. A real physical system cannot produce such an abnormal response, hence these impulses are certainly unrelated to the bearing. Note that the filtered signal in Fig. 12(a) contains the same type of impulses.

The selected region is sufficient ($_{rlen}=32.2\%$) and is thus analysed using the envelope technique [1],[27] and [28]. Envelope analysis (also known as the high-frequency resonance technique) is an established diagnostic method in which a specific frequency band is demodulated in order to shift modulation effects at high frequency into the low-frequency range. To avoid aliasing and spurious harmonics [28], the envelope signal is obtained using the Hilbert transform and is squared before computing its spectrum by the DFT. The demodulation band is chosen from 5500 to 6500 Hz, yielding the squared envelope in Fig. 16(f). The squared envelope spectrum contains several harmonics of 84 Hz (Fig.16 (g)), which indicates a damaged bearing cage since the theoretical FTF is 82 Hz (Table 1). The presence of the fault is also indicated by kurtosis of the squared envelope $_{\beta2}(e^2)=5.21$, while the raw signal and the selected region have a low kurtosis value of 2.61 and 2.35, respectively.

Signals from washing machines

Fig. 17(a) shows a typical velocity signal measured on a drum of a WM. Only a detail of the steady-state waveform is shown here; the complete measurement is depicted in Fig. 18(a). The waveform resembles a noisy sinusoid and is practically the same as the signals with speckle noise in [6] and [7]. The frequency of the sinusoid is $f_0 = 15.1\,Hz$, which corresponds to the speed of the drum and represents the dominant frequency component of the signal.

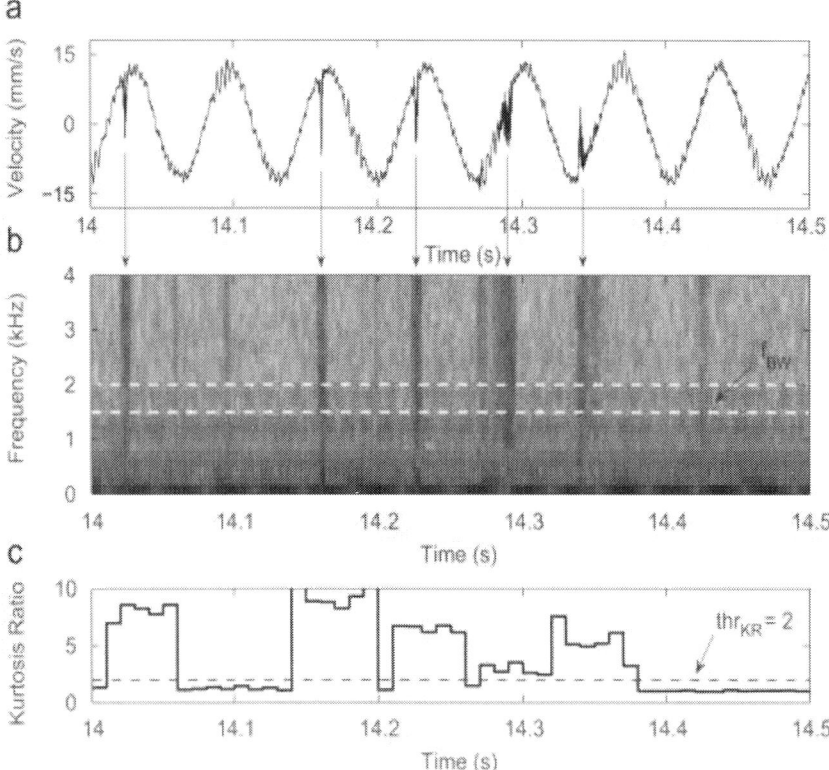

Figure 17. Speckle noise in the signal from a faulty WM: (a) raw signal with amplitude drop-outs, (b) selecting the filtered band using the spectrogram, (c) $KR_{(xf)}$.

It is seen that speckle noise generates "spikes" on the waveform of the velocity signal. The direction of these spikes is always towards zero, which implies that the speckle noise causes drop-outs of the signal amplitude. Therefore, the character of random impulses differs from those in Fig. 5, where the signal amplitude was increased and resulted in outliers. Nevertheless, both types of random impulses are narrowly localised in the time domain, hence their amplitude spectrum is spread over a wide range of frequencies (which causes the vertical lines in the spectrogram). This explains why the speckle noise is a high-level broadband noise, a property to be further illustrated in Example 5.

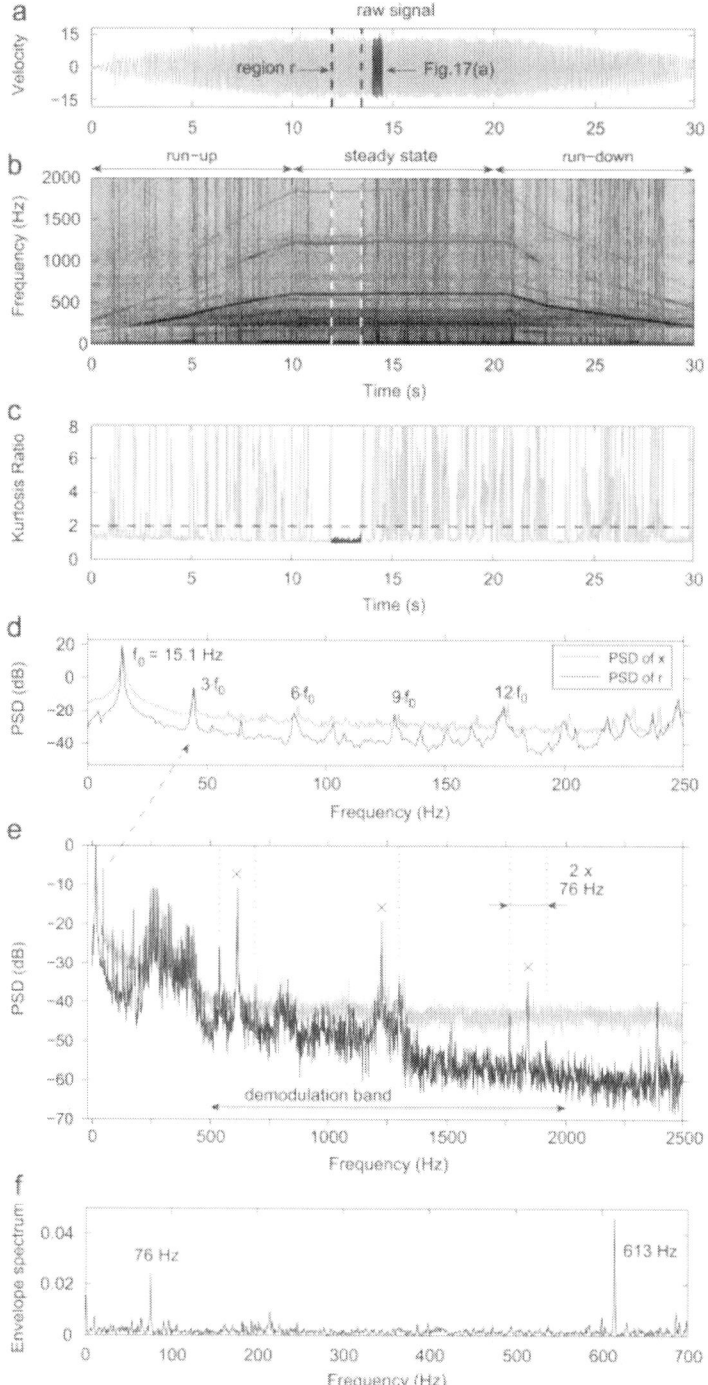

Figure 18. Example 5—analysis of the first faulty washing machine.

Selecting the filtered band $_{fBW}$ is easier than in Section 4.2 and can be achieved simply by observing the spectrogram in Fig. 17(b). An example band from 1500 to 2000 Hz yields satisfactory results, but choosing a higher range would work equally well since the impulses are visible up to 4 kHz. This is in agreement with signals from electric motors, for which the impulses propagate even up to 12 kHz (e.g. Fig. 16(b)). This extreme bandwidth is a consequence of the non-physical origin of the speckle noise. For this reason, it seems always to be possible to find some (high-frequency) band, where the speckle noise prevails over other (physical) vibration sources. Moreover, the method can operate in narrow bandwidths (i.e. the example here as opposed to 5–10 kHz for electric motors), which would be an advantage for applications where a suitable band is harder to find.

Fig. 17(c) shows the segment KR computed with $f_{BW} = (1500, 2000)$ Hz $_{wlen}=1000, _{wovp}=80\%$ and k=5%. The window length $_{wlen}$ is shorter than the average signal period of 1325 samples $(f_s/f_0 = 20\,000/15.1 \approx 1325)$, hence the window step of 200 samples provides sufficient time resolution for distinguishing between the periods with and without speckle noise. The default threshold $thr_{KR}=2$ seems suitable also for signals from WMs.

Example 5—faulty motor

Fig. 18(a) depicts the signal from a faulty WM, a detail of which was shown in Fig. 17(a). The spectrogram in Fig. 18(b) reveals that the signal is very noisy during the run-up and run-down, thus making the transient-state analysis particularly difficult (e.g. order tracking cannot be performed). Fig. 18(c) shows the segment KR which proves to be a versatile tool for localising undistorted segments. The indicators in Table 4 are computed only from the steady-state part of the measurement, which represents the raw signal x. In other words, $_{nerr}=71$ is counted between 10 and 20 s and $_{rlen}=14.7\%$ thus corresponds to 1.47 s. The region r contains about 22 rotations of the drum and 300 rotations of the motor.

Figs. 18(d) and (e) show the power spectral densities of x and r, which imply that speckle noise is an additive measurement noise. It is seen that avoiding the noise enhances the PSD in both the low- and high-frequency range. The former range (0–250 Hz) contains mainly the harmonics of f_0, partially covered by several dB of speckle noise. In the latter range (1500–2500 Hz), the noise increases the overall PSD level by almost 20 dB and completely masks the peak at 1839 Hz and its sidebands. For this reason, speckle noise can be regarded as a special case of random masking (as opposed to discrete masking typically represented by gearmesh-frequencies [21], [24] and [28]). When the noise is reduced, the enhanced

PSD exhibits peaks at 613, 1226 and 1839 Hz, accompanied by sidebands of 76 Hz. These peaks are located in the demodulation band $(500, 2000)$ Hz, hence the frequency of 613 Hz dominates the squared envelope spectrum in Fig. 18(f). This frequency is close to $3f_r = 615$ Hz and provides an evidence of a motor-related fault. Indeed, this feature is present in all measurements with faulty motors and is absent in signals from good WMs. The envelope spectrum also shows the frequency of 76 Hz, which indicates modulation at the cage frequency (FTF, Table 1). This suggests that some rollers might be more worn than the others. Finally, note that although the LDV signal is measured on the surface of the drum, frequency components related to the motor and its bearings can be extracted due to sufficient dynamic range of the laser vibrometer (>90 dB [14]).

Example 6—severe fault

Fig. 19(b) depicts a velocity signal from another WM with a faulty motor. The signal waveform is shown in Fig. 19(a), where the oscillations caused by a severe mechanical fault are easily seen. Amplitude drop-outs can be observed as in Fig. 17(a), suggesting that speckle noise occurs independently of the machine condition. The same phenomenon was observed in all measurements, including those from good WMs.

Thresholding of $\mathbf{KR}(x_f)$ gives a similar result to that in the previous example, which is expressed by comparable values of the indicators (Table 4). Note that there is another region during the run-down (26–28 s) which is longer than the selected steady-state region. The PSD of r in Fig. 19(e) is dominated by a family of significant peaks with a spacing of 201 Hz, indicating a strong (amplitude) modulation at this frequency. The frequency of 201 Hz corresponds to the average speed of the motor which operates at a lower rate than the nominal speed of 205 Hz. The demodulation process shifts the sideband spacing into the low-frequency range, hence the peak at 201 Hz can be detected directly in the squared envelope spectrum (Fig. 19(f)). The presence of rotor speed and its low harmonics indicates a mechanical looseness, probably related to a released belt or clamping screws of the electric motor (common defects in WMs [3]). The third harmonic (605 Hz) is larger than the fundamental, which is in agreement with Fig. 18(f) where the frequency of 3_{fr} is also excited.

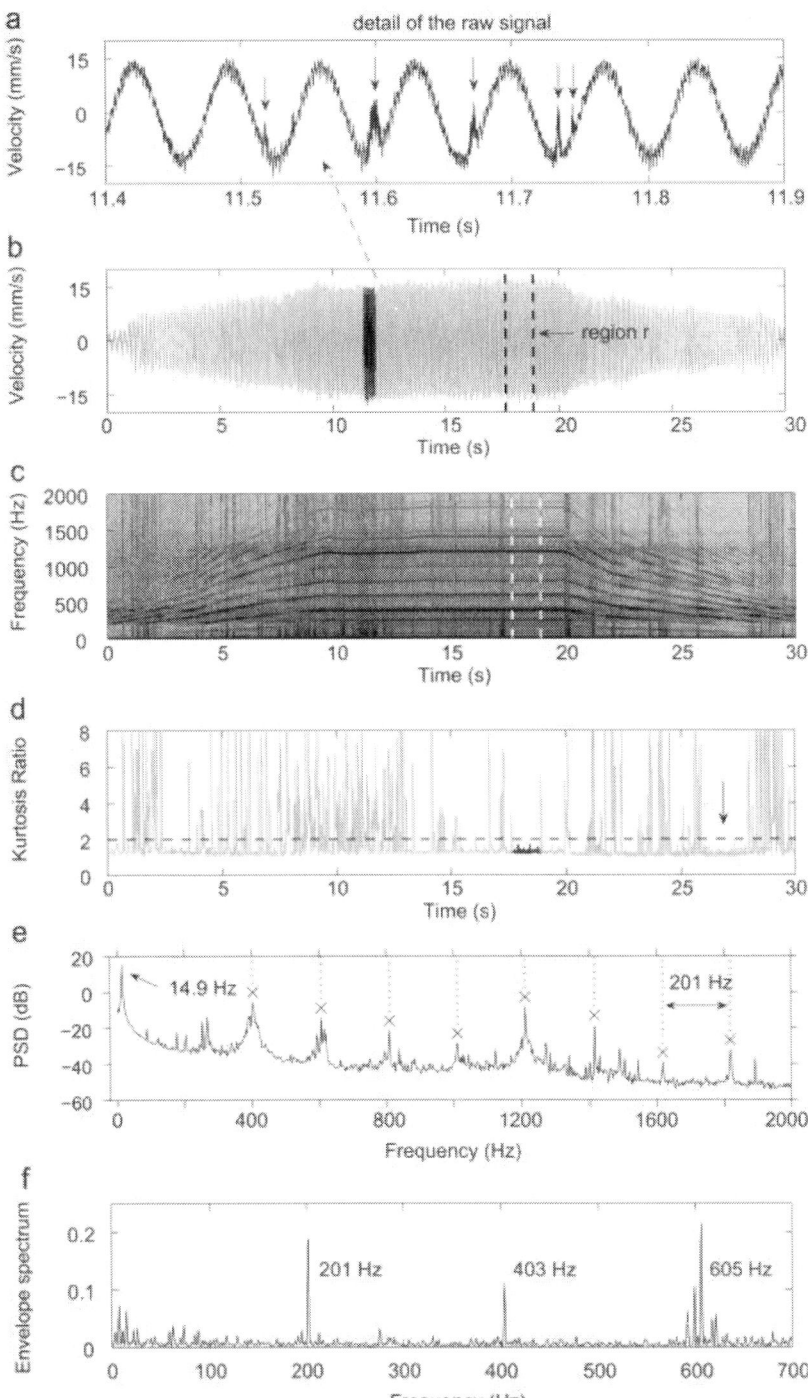

Figure 19. Example 6—analysis of the second faulty washing machine.

CONCLUSIONS

This paper presents a statistical method for avoiding speckle noise in laser Doppler vibrometry. The presence of speckle noise is detected using a new scalar indicator kurtosis ratio (KR), defined as a ratio of kurtosis and a robust estimate of kurtosis. As a result, the KR indicates the amount of outlying samples in the data, and is thus efficient in detection of random impulses generated by the speckle noise. The value of the KR depends on a single parameter (outlier removal level) which determines the amplitude of two automatic thresholds based on percentiles. Therefore, the detector has a clear physical meaning and is easy to implement in Matlab or Lab VIEW.

Since measurements cannot be repeated during the on-line quality control, an algorithm for selecting an undistorted portion of a signal has been developed. The algorithm is fast and simple since all computations are performed in the time domain. Reliable results are obtained owing to three key features. First, the KR is calculated by removing a fixed percentage of outlying samples, which guarantees that speckle noise is detected independently of the impulse peak amplitudes or their total number. Second, sensitivity of the KR is significantly improved by band-pass filtering which enhances the random impulses hidden in the raw vibration signal. Third, the KR is applied on overlapping segments of the filtered signal, resulting in the segment KR. This function is sensitive to spectral changes in the STFT and is efficient in localising signal segments unaffected by the speckle noise. In addition, each algorithm parameter is discussed in detail and instructions for optimisation are provided. Only two parameters must be adjusted by the user, since the recommended window overlap, the threshold for the KR and the outlier removal level should be appropriate for most applications.

Simulated results show that the method can reliably distinguish between the random impulses of speckle noise and periodic impulses of bearing faults, provided that the bearing impulses are sufficiently buried in the background noise. This is confirmed by experimental results showing that the presence of mechanical faults (detected by envelope analysis) does not significantly interfere with detection of speckle noise. The versatility of the algorithm is demonstrated on two types of LDV signals: from electric motors and washing machines. It is shown that speckle noise has properties of additive broadband noise and that it can be avoided even in severely distorted signals, hence the improvement of the SNR is significant. The method can thus play an important role in LDV-based diagnostic systems, since validation of measured data should be always performed prior to fault detection, particularly in harsh conditions of the production lines.

ACKNOWLEDGEMENTS

This work has been supported by the GA ČR Grant No. 102/03/H085 "Biological and Speech Signal Modelling" and the research program no. MSM6840770015 "Research of Methods and Systems for Measurement of Physical Quantities and Measured Data Processing" of the CTU in Prague sponsored by the Ministry of Education, Youth and Sports of the Czech Republic. A part of this work has been carried out during the Practicum Program of J. Vass at the University of New South Wales (UNSW), Sydney, Australia. J. Vass wishes to acknowledge the CTU and Hlávka's foundation for supporting his research at the UNSW. He also wishes to thank the following friends and colleagues from the UNSW who have commented on earlier versions of this paper: H. Endo, D. Hanson, N. Sawalhi, M. Skeen, A. Hespe and N.J. Kessissoglou. The authors are also grateful for the comments of the reviewers.

REFERENCES

1. N. Tandon, A. Choudhury, A review of vibration and acoustic measurement methods for the detection of defects in rolling element bearings, Tribology International 32 (8) (1999) 469–480.
2. M.E.H. Benbouzid, A review of induction motors signature analysis as a medium for faults detection, IEEE Transactions on Industrial Electronics 47 (5) (2000) 984–993. ARTICLE IN PRESS 670 J. Vass et al. / Mechanical Systems and Signal Processing 22 (2008) 647–671
3. N. Paone, L. Scalise, G. Stavrakakis, A. Pouliezos, Fault detection for quality control of house-hold appliances by non-invasive laser Doppler technique and likelihood classifier, Measurement 25 (1999) 237–247.
4. C. Cristalli, N. Paone, R.M. Rodrı´guez, Mechanical fault detection of electric motors by laser vibrometer and accelerometer measurements, Mechanical Systems and Signal Processing 20 (6) (2006) 1350–1361.
5. J. Vass, C. Cristalli, Bearing fault detection for on-line quality control of electric motors, in: Proceedings of the 10th IMEKO TC10 International Conference on Technical Diagnostics, Budapest, Hungary, 9–10 June 2005, pp. 93–97.
6. M. Gasparetti, G.M. Revel, The influence of operating conditions on the accuracy of in-plane laser Doppler velocimetry measurements, Measurement 26 (3) (1999) 207–220.
7. E.P. Tomasini, G.M. Revel, P. Castellini, Laser based measurements, Encyclopedia of Vibration, Academic Press, London, UK, 2001, pp. 699–710, doi:10.1006/rwvb.2001.0152.

8. P. Castellini, M. Martarelli, E.P. Tomasini, Laser Doppler vibrometry: development of advanced solutions answering to technology's needs, Mechanical Systems and Signal Processing 20 (6) (2006) 1265–1285.
9. S.J. Rothberg, J.R. Baker, N.A. Halliwell, Laser vibrometry: pseudo-vibrations, Journal of Sound and Vibration 135 (3) (1989) 516–522.
10. N.A. Halliwell, The laser torsional vibrometer: a step forward in rotating machinery diagnostics, Journal of Sound and Vibration 190 (3) (1996) 399–418.
11. M. Denman, N.A. Halliwell, S.J. Rothberg, Speckle noise reduction in laser vibrometry: experimental and numerical optimisation, Proceedings of the SPIE 2868 (1996) 12–22.
12. B.J. Halkon, S.J. Rothberg, Vibration measurements using continuous scanning laser vibrometry: advanced aspects in rotor applications, Mechanical Systems and Signal Processing 20 (6) (2006) 1286–1299.
13. M. Johansmann, G. Siegmund, M. Pineda, Targeting the limits of laser Doppler vibrometry, in: Proceedings of the IDEMA 2005, Tokyo, Japan, 2005, pp. 1–12.
14. Polytec, IVS-300 Industrial Vibration Sensor—Technical specifications, 2005.
15. E.W. Weisstein, Kurtosis, MathWorld—A Wolfram Web Resource hhttp://mathworld.wolfram.com/Kurtosis.htmli.
16. The MathWorks, Inc., Statistics Toolbox User's Guide (Version 6.0), 2007.
17. D. Dyer, R.M. Stewart, Detection of rolling element bearing damage by statistical vibration analysis, Journal of Mechanical Design 100 (2) (1978) 229–235.
18. H.R. Martin, F. Honarvar, Application of statistical moments to bearing failure detection, Applied Acoustics 44 (1) (1995) 67–77.
19. C. Pachaud, R. Salvetat, C. Fray, Crest factor and kurtosis contributions to identify defects inducing periodical impulsive forces, Mechanical Systems and Signal Processing 11 (6) (1997) 903–916.
20. J.P. Dron, F. Bolaers, L. Rasolofondraibe, Improvement of the sensitivity of the scalar indicators (crest factor, kurtosis) using a denoising method by spectral subtraction: application to the detection of defects in ball bearings, Journal of Sound and Vibration 270 (1–2) (2004) 61–73.
21. J. Antoni, R.B. Randall, The spectral kurtosis: application to the vibratory surveillance and diagnostics of rotating machines, Mechanical Systems and Signal Processing 20 (2) (2006) 308–331.
22. A. Hyva¨rinen, Sparse code shrinkage: denoising by nonlinear maximum likelihood estimation, Neural Computation 11 (7) (1999) 1739–1768.
23. J. Lin, M.J. Zuo, K.R. Fyfe, Mechanical fault detection based on the wavelet de-noising technique, ASME Journal of Vibration and Acoustics 126 (1) (2004) 9–16.

24. W. Wang, A.K. Wong, Autoregressive model-based gear fault diagnosis, ASME Journal of Vibration and Acoustics (124) (2002) 172–179.
25. T. Sarama"ki, Finite impulse response filter design, in: S.K. Mitra, J.F. Kaiser (Eds.), Handbook for Digital Signal Processing, WileyInterscience, New York, USA, 1993 (Chapter 4).
26. R.F. Dwyer, Use of the kurtosis statistic in the frequency domain as an aid in detecting random signals, IEEE Journal of Oceanic Engineering OE- 9 (2) (1984) 85–92.
27. N. Sawalhi, R.B. Randall, H. Endo, The enhancement of fault detection and diagnosis in rolling element bearings using minimum entropy deconvolution combined with spectral kurtosis, Mechanical Systems and Signal Processing 21 (2007) 2616–2633, doi:10.1016/ j.ymssp.2006.12.002.
28. D. Ho, R.B. Randall, Optimisation of bearing diagnostic techniques using simulated and actual bearing fault signals, Mechanical Systems and Signal Processing 14 (5) (2000) 763–788.
29. J. Vass, C. Cristalli, Optimization of Morlet wavelet for mechanical fault diagnosis, in: Proceedings of the 12th International Congress on Sound and Vibration (ICSV12), Lisbon, Portugal, 11–14 July 2005 hhttp://amber.feld.cvut.cz/user/vass/papersi.

CITATION

J. Vass, R. Šmíd, R.B. Randall, P. Sovka, C. Cristalli, B. Torcianti, Avoidance of speckle noise in laser vibrometry by the use of kurtosis ratio: Application to mechanical fault diagnostics, Mechanical Systems and Signal Processing, Volume 22, Issue 3, April 2008, Pages 647-671, ISSN 0888-3270, http://dx.doi.org/10.1016/j. ymssp.2007.08.008.

CHAPTER 6

Evaluating the Integrity of Pressure Pipelines by Fracture Mechanics

ĽubomírGajdoš and Martin Šperl

Institute of Theoretical and Applied Mechanics, Academy of Sciences of the Czech Republic, Czech Republic

INTRODUCTION

Large engineering structures made with the use of sophisticated technology often include material defects and geometrical imperfections. These defects or imperfections do not exert their influence on the initial behaviour of structures designed in accordance with standard rules. Under the action of loading varying in time, however, they can reveal themselves in long-term operation by the initiation and growth of a fatigue crack from a defect root. Similarly, stress corrosion (SC) cracks can develop in a structure when there is an initial stress concentrator and the structure is exposed to both mechanical stress and a corrosion medium. A condition for the growth of a small fatigue crack is that the level of cyclic stress should be above the limit value given by barriers existing in a steel, and a condition for the growth of SC cracks is that the stress is greater than a certain limit value for a specific corrosion medium. It is important to pay due attention to the behaviour of cracks under various gas pipeline loading conditions in different environments, and to the influence of these conditions on the residual strength and life of the gas pipeline. The existence of a crack in the wall of a high-pressure gas pipeline mostly implies a shortened remaining period of reliable operation.

THEORETICAL TREATMENT OF CRACKS IN PIPES

At the present time, the manufacturing stage of pipes for gas pipelines includes sufficient flaw detection measures, and only products free of detectable material flaws are dispatched for operation. However, there are defects that are not revealed by the required inspection, and which manifest themselves during heavy-duty operation. The most dangerous defect is the occurrence of cracks – these are due to material defects that are difficult to reveal by a standard optical inspection. If the cracks are deep, and spread to a large extent, they can pose a threat to the pipeline operation. Using fracture mechanics it is possible to evaluate the threat that crack-like defects can pose to the pipeline wall, depending on whether a brittle, quasi-brittle or ductile material is involved. A model description of crack-containing systems, which relies on the stress intensity factor, K, can be used for brittle and quasi-brittle fracture, and also for subcritical fatigue growth, corrosion fatigue, and stress corrosion. In these cases, the surface crack is usually located in the field of one of the membrane tensile stress components, or in the field of bending stress, or in a combination of both. The extent of the plastic zone at the crack tip is small in comparison with the dimensions of the crack and the pipeline.

If the gas pipeline is made of a high-toughness material, the plastic strains become extensive before the crack reaches instability. Hence, some elasto-plastic fracture mechanics parameter, such as the J–integral or crack opening displacement, or a two-criterion method, should be employed to assess the threat that the crack poses to the pipeline wall. Although cracks of various directions may occur in the pipe wall, we will consider here only longitudinal cracks, because they are subjected to the biggest stress (hoop stress) in the pipe wall, and they are therefore the most dangerous (when we are considering the parent metal).

Stress Intensity Factor for a Longitudinal Through Crack inthe Pipe Wall

The first theoretical solution to the problem of establishing the stress intensity factor for a long cylindrical pipe with a longitudinal through crack under internal pressure was reported by Folias (Folias, 1969) and by Erdogan with Kibler (Erdogan &Kibler, 1969). They managed to show that the problem was analogous to that of a wide plane plate with a through crack. The only adjustment needed for transition to a pipe was to introduce a correction factor to multiply the solution for the plane plate. This factor, frequently referred to as the Folias correction factor and designated by symbol M_T, is only a function of the ratio $\lambda = c/\sqrt{Rt}$, where c is the crack half-length, R is the pipe mean radius, and t is the pipe wall thickness, and thus it depends only on the geometrical parameters of the crack and the pipe (Fig. 1).

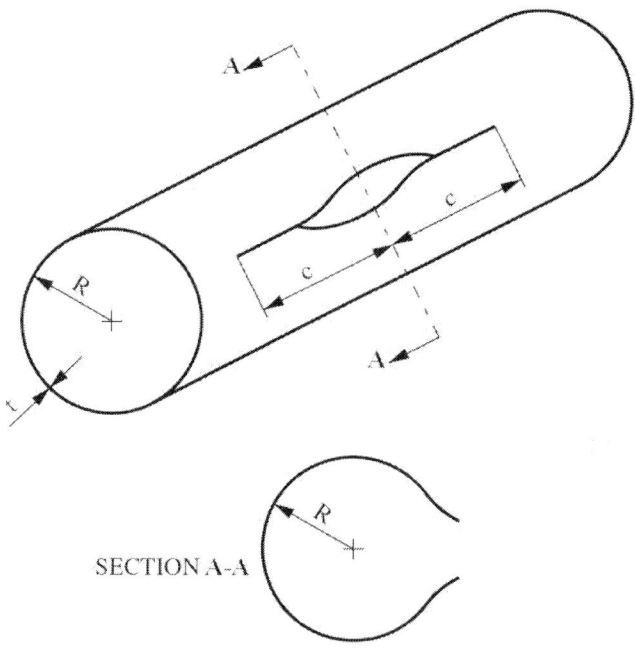

SECTION A-A

Figure 1. Deformation of a pressurized pipe in the vicinity of a longitudinal through crack

Several relations have been reported for determining the Folias correction factor. The following are the most frequently used at the present time:
The Folias relation (Folias, 1970):

$$M_T = \sqrt{1 + 1.255\lambda^2 - 0.0135\lambda^4} \tag{1}$$

The Erdogan et al. relation (Erdogan et al., 1977):

$$M_T = 0.6 + 0.5\lambda + 0.4\exp(-1.25\lambda) \qquad \text{for} \quad \omega < 5$$
$$M_T = 1.761(\lambda - 1.9)^{0.5} \qquad\qquad \text{for} \quad \omega \geq 5$$

where

$$\omega = \sqrt[4]{12(1 - \nu^2)\lambda} \tag{3}$$

with ν denoting Poisson's ratio

The following relation is the simplest:

$$M_T = \sqrt{1 + 1.61\lambda^2} \tag{4}$$

However, its validity is limited by the value $\lambda < 1$.

If c is the half-length of a longitudinal through crack in the pipe, then the stress intensity factor of such a crack simply reads

$$K_I = M_T \sigma_\varphi \sqrt{\pi c} \tag{5}$$

Where

$\sigma\varphi = \pi D/2t$ is the hoop stress (D and t denoting pipe diameter and wall thickness, respectively), and MT is the Folias correction factor

Stress Intensity Factor for a Longitudinal Part-Through Crack

Various methods are used for analyzing the problem of longitudinal semi-elliptical surface cracks in the wall of cylindrical shells (Fig. 2). As a 3D asymptotic solution to the stress intensity factor is virtually involved, the possibilities offered by accurate analytical procedures are confined to infinite or semi-infinite bodies. Solutions appropriate for finite bodies call for the application of approximate methods, such as the finite element method and the method of boundary integral equations, or various alternative methods (e.g. the weight function method).

The first solutions for semi-elliptical surface cracks in a plate subjected to uniaxial tension or steady bending were derived from solutions for an elliptical plane crack in an infinite 3D body. In order to account for the finite thickness of a body and the plastic zone at the crack tip, correction factors were introduced for the "front" surface and the "rear" surface of the body and for the plastic region at the crack tip (Shah & Kobayashi, 1973). However, solutions by different authors often showed rather considerable disagreement. Scott and Thorpe (Scott & Thorpe, 1981) therefore tested the accuracy of the solutions presented by various authors by measuring changes in the shape of a crack throughout its fatigue growth. They concluded that the best engineering estimation of the stress intensity factor for a part-through crack in a plate was provided by Newman's solution (Newman, 1973). An adjusted form of this solution for a thin-walled shell is given by:

$$K_I = \left[M_F + \left(E_{(k)} \sqrt{c/a} - M_F \right) \left(\tfrac{a}{t} \right)^s \right] \frac{\sigma_\varphi \sqrt{\pi a}}{E_{(k)}} M_{TM} \tag{6}$$

Where

MF is the function depending on the crack geometry (on the ratio a/c)

$$E_{(k)} = \int_0^{\pi/2} \sqrt{1 - k^2 \sin^2 \theta} \, d\theta \text{ is an elliptical integral of the second kind, } k \text{ being } \sqrt{1 - \left(\tfrac{a}{c} \right)^2}$$

s is the function depending on the crack geometry (the ratio a/c) and the relative crack depth (the ratio a/t)

$$M_{TM} = \frac{\left(1 - \frac{a/t}{M_T} \right)}{(1 - a/t)} \text{ is the correction factor for the curvature of a cylindrical}$$

shell and for an increase in stress owing to radial strains in the vicinity of the crack root.

In the last relationship, M_T is the Folias correction factor, determined by any of the relations (1) – (3). The functions M_F and s differ in form for the lowest point of the crack tip (point A in Fig. 2) and for the crack mouth on the surface of the cylindrical shell (point B in Fig. 2).

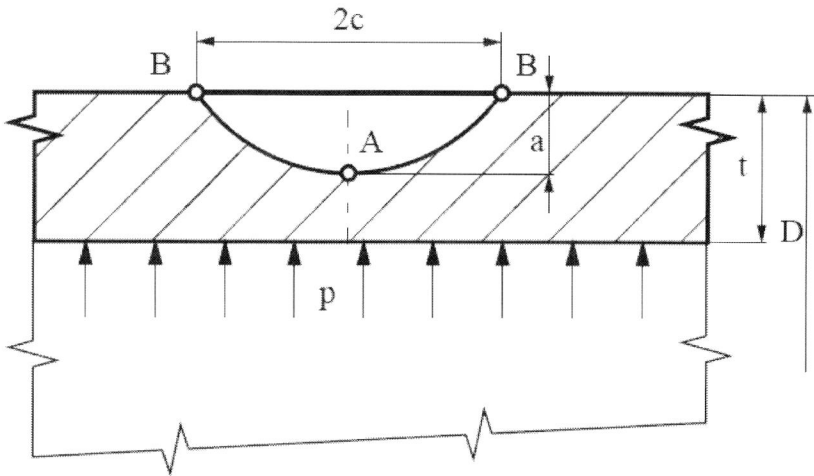

Figure 2. External longitudinal semi-elliptical crack in the wall of a cylindrical shell

Engineering Methods for Determining the J Integral
The FC Method

This method was proposed as the J_s method in Addendum A16 of the French nuclear code (RCC-MR, 1985). It stems from the second option for describing the transition state between ideally elastic and fully plastic behavior of a material, i.e. from the function $f_2(L_r)$ of the R6 method (Milne et al., 1986). This function takes the form:

$$f_2(L_r) = \left(\frac{E\varepsilon_{ref}}{L_r R_e} + \frac{L_r^3 R_e}{2 E\varepsilon_{ref}} \right)^{-1/2} \tag{7}$$

Where

$Lr = \sigma/\sigma L$ (σ – applied stress, σL – stress at the limit load)

R_e is the yield stress

E is Young's modulus

ε_{ref} is the reference strain corresponding to the reference (nominal) stress σ_{ref}

If we identify function $f_2(L_r)$ with function $f_3(L_r) = \left(\frac{J}{J_e} \right)^{-\frac{1}{2}}$ and express L_r as σ_{ref}/R_e and the elastic J integral Je as K^2/E', where E' =E for plane stress state and E' =E / (1−v²) for plane strain state, we have:

$$J = \frac{K^2}{E'} \left(\frac{E.\varepsilon_{ref}}{\sigma_{ref}} + \frac{\sigma_{ref}^3}{2E.R_e^2.\varepsilon_{ref}} \right) \tag{8}$$

The stress σref in the above equation is a nominal stress – i.e. a stress acting in the plane where the crack occurs. Taking into consideration the description of the stress-strain dependence by the Ramberg-Osgood relation (8) and adjusting Eq. (7), we obtain the J-integral in the form (9).

$$\frac{\varepsilon}{\varepsilon_0} = \frac{\sigma}{\sigma_0} + \alpha \left(\frac{\sigma}{\sigma_0} \right)^n \tag{9}$$

$$J = \frac{K^2}{E'}\left[A + \frac{0.5(\sigma/\sigma_0)^2}{A}\right] \tag{10}$$

where

$$A = 1 + \alpha\left(\frac{\sigma}{\sigma_0}\right)^{n-1} \tag{11}$$

In the above equations the stress σ_0 can be substituted by the yield stress Re ; $\varepsilon_0 = \sigma_0/E$; α, n – material constants.

As a pipeline is a body of finite dimensions, stress σ in Eqs. (9) and (10) is a nominal stress – i.e. a stress acting in the plane where the crack occurs. Referring to the R6 method (Milne et al., 1986), this stress for a pipe containing a longitudinal part-through thickness crack may be written as:

$$\sigma = \frac{\sigma_\phi}{1 - \frac{\pi a c}{2t(t+2c)}} \tag{12}$$

In eq. (11), $\sigma_\phi = \frac{pD}{2t}$ is the hoop stress, and the meaning of the symbols a, c, and t is clear from Fig. 2.

GS Method

The GS method was derived by Gajdoš and Srnec (Gajdoš&Srnec, 1994) on the basis of the limit transition of the J-integral, formally expressed for a semi-circular notch, to a crack, with the variation of the strain energy density along the notch circumference being approximated by the third power of the cosine function of the polar angle. If the stress-strain dependence is further expressed by the Ramberg-Osgood relation (8), with $\varepsilon_0 = \sigma_0 / E$, (α, n – material constants), we can arrive at Eq. (12)

$$J = \frac{K^2}{E'}\left[1 + \frac{2\alpha n}{(n+1)}\left(\frac{\sigma}{\sigma_0}\right)^{n-1}\right] \tag{13}$$

where σ is the nominal stress in the reduced cross-section of a body. For a pipe containing a longitudinal part-through thickness crack it may be determined by relation (11).

CONSIDERATION OF THE CONSTRAINT

As mentioned above, the situation existing at the crack tip in conditions of small-scale yielding can be characterized by a single fracture parameter (e.g. K, J or δ). This parameter can be used as a fracture criterion, independent of geometry. However, single-parameter fracture mechanics fails in cases of developed plasticity, where fracture toughness is a function not only of the material, but also of the dimensions and the geometry of the specimen. It is well known from the theory of fracture mechanics that for small-scale yielding the maximum stress existing at the crack tip in a non-hardening material is about $3\sigma_0$, where σ_0 is the yield stress. Single-parameter fracture mechanics apparently does not apply to non-hardening materials under fully plastic conditions, because the stress and strain fields in the vicinity of the crack tip are affected by configurations of both the body and the crack. The situation is more favourable in hardening materials, where single-parameter fracture mechanics may approximately apply also for the developed plasticity, provided that the body maintains a high level of stress triaxiality.

The reported experimental studies suggest that the configuration of the specimen and the crack (the crack depth and the specimen dimensions, in particular) affect the fracture toughness in a brittle state. However, the fact that this configuration can also influence the R-curve of ductile materials is not so well known.

Generally, the bigger the dimension of the crack, the smaller the resistance of the material to fracture will be. The R-curve obtained on specimens with rather long cracks is, as a rule, below the R-curve obtained on specimens with rather short cracks. For this reason, standards require that the relative crack lengths be within a comparatively narrow range of values for valid values of fracture toughness J_{in}.

The J – Q Theory
Some researchers dealing with fracture mechanics tried to extend the theory of fracture mechanics beyond the boundaries of the assumptions of single-parameter fracture mechanics, introducing other parameters to provide a more accurate characterization of conditions at the crack tip. One

of the parameters is the so-called T-stress, which is a uniform stress acting axially (in the direction of the x-axis) in front of the crack tip in an isotropic elastic material loaded by the first mode, i.e. the opening mode, of the load. In this case, the stress field in front of the crack tip may be written as:

$$\sigma_{ij} = \frac{K_I}{\sqrt{2\pi r}} f_{ij}(\Theta) + T\delta_{1i}\delta_{1j} \tag{14}$$

The elastic T-stress heavily affects the shape of the plastic zone and the stress deep in this zone. T-stress values are linked with the stress biaxiality ratio, β, defined as

$$\beta = \frac{T\sqrt{\pi a}}{K_I} \tag{15}$$

It can be mentioned by way of illustration that the stress biaxiality ratio β equals -1 for a through crack in an infinite plate loaded by a normal stress applied far away from the crack plane. By implication, this remote stress, σ, induces a T-stress in the direction of the x-axis, whose magnitude is -σ. In an elastic case, positive values of the T-stress generally lead to a high constraint under fully elastic conditions, whereas a geometry with a negative T-stress leads to a rapid drop in the constraint as the load rises. For different geometries, the stress biaxiality ratio β can be used as a qualitative index for a relative constraint at the crack tip.

The so-called J – Q theory provides another approach to the extension of single-parameter fracture mechanics beyond the conditions of its validity. This theory aims to describe the stress field at the crack tip deep in the plastic zone. It is a well-known fact that if the small-strain theory is used, the stress field at the crack tip in the plastic zone can be described by a power series, in which the so-called HRR solution is the leading term (Hutchinson, 1968), (Rice &Rosengren, 1968). The other terms of higher magnitudes, when summed up, provide a difference stress field, which approximately corresponds to a uniform hydrostatic shift of the stress field in front of the crack tip. It has become customary to designate the amplitude of this approximate difference stress field with letter Q, according to its authors O'Dowd and Shih (O'Dowd & Shih, 1991). O'Dowd and Shih defined the Q parameter as:

$$Q \equiv \frac{\sigma_{yy} - (\sigma_{yy})_{HRR or} T{-}0}{\sigma_0} \tag{16}$$

$$\text{for } \Theta = 0 \text{ and } \frac{r\sigma_0}{J} = 2$$

The parameter is equal to zero ($Q = 0$) under small-scale yielding conditions, but it acquires negative values as the load (and in consequence the strain) grows. Classical single-parameter fracture mechanics assumes that fracture toughness is a material constant. However, the J-Q theory suggests that the critical value of the J-integral for a given material depends on the Q parameter – i.e. $J_c = J_c(Q)$– and that fracture toughness is thus not some single-value quantity, but rather a function that defines the critical values of the J-integral and the Q parameter (Shih et al., 1993). Although the relation between critical J-integral values and the Q parameter shows a considerable scatter, the critical value of the J-integral tends in general to drop as the Q parameter increases in value.

The theory of single-parameter fracture mechanics assumes that the fracture toughness values obtained on laboratory specimens can be applied to a body. However, two-parameter approaches, such as the J-Q theory, reveal that the specimen must be tested at the same constraint as that of the body with a crack. In other words, the two geometries must have the same Q value at the moment of fracture, so that the corresponding critical values of the J-integral, J_{cr}, will be equal to each other. Since J_{cr} values are often scattered to a large extent, we cannot make a clear-cut prediction of this quantity. It is only possible to predict a certain range of plausible J_{cr} values for a given body or structure.

It should also be noted that the J-Q approach is only descriptive, and not predictive. This implies that the Q parameter quantifies the constraint at the crack tip, without providing any indication of the particular influence of the constraint on the fracture toughness. Two-parameter theories cannot be strictly correct as far as their universality is concerned, because they assume two degrees of freedom. Recent research into the influence of the constraint at the crack tip on fracture toughness indicates that geometries with a low constraint can in many cases be judged by a two-parameter theory, and geometries with a high constraint can be judged by a single-parameter theory (Ainsworth & O'Dowd, 1995).

Plastic Constraint Factor on Yielding

A simple procedure based on the use of the so-called plastic constraint factor on yielding, C, can be applied to determine the fracture conditions in a thin-walled pressure pipeline. The factor is given by the ratio of the stress needed to obtain plastic macrostrains under constraint conditions to the yield stress at a homogeneous uniaxial state of stress (Gajdoš et al., 2004). The C factor can be expressed by the relation (16)

$$C = \frac{\sigma_1}{\sigma_{HMH}} \tag{17}$$

where σHMH, the Huber-Mises-Hencky stress, is put equal to the yield stress.

Let us now consider the state of stress at the crack tip in a thick-walled body, where the stress perpendicular to the crack plane, σ_1, and the stress in the direction of the crack, σ_2, are equal, and the stress in the direction of the thickness of the body, σ_3, is governed by the expression $\sigma_3 = v(\sigma_1 + \sigma_2)$. Then, based on the HMH criterion and assumed elastic conditions ($v \cong 0.33$), the plastic constraint factor $C \approx 3$. If the stress in the thickness direction, σ_3, falls between $2v\sigma_1$ and zero (thin-walled body), the value of the plastic constraint factor will range between $C = 1$ and $C = 3$. This data can be used to assess the fracture conditions in gas pipelines with surface part-through cracks, employing a C-factor which has to be experimentally determined. After the C factor has been determined, the value of $C\sigma_0$ would be used instead of the yield stress σ_0 in relations for calculating the J-integral. The C factor was experimentally investigated at the Institute of Theoretical and Applied Mechanics of the Academy of Sciences of the Czech Republic in the framework of a broader research project on the reliability and operational safety of high pressure gas pipelines. Fracture conditions were investigated on five pipe bodies, made of steels X52, X65 and X70, with cycling-induced cracks. Data on the pipe bodies that were used, the cracks in the walls, and the mechanical and fracture-mechanical material properties of the bodies are given in Table 1.

The rows in the table show the following data (top to bottom): body diameter D, body wall thickness t, half-length of a longitudinal part-through crack c, crack depth a, relative crack depth a/t, aspect ratio a/c of a semi-elliptical crack, fracture pressure p, ratio of fracture pressure p and pressure $p_{0.2}$ corresponding to the hoop stress at the yield stress, yield stress in the circumferential direction of the body σ_0, Ramberg-Osgood constant

Table 1. Summary of data on the assessment of the fracture behaviour of model pipe bodies

Material	X 52	X 65	X 65	X 70	X 70
D (mm)	820	820	820	1018	1018
t (mm)	10.2	10.7	10.6	11.7	11.7
c (mm)	50	100	100	127	115
a (mm)	7	7.7	7	6.7	7.1
a/t	0.686	0.72	0.66	0.573	0.607
a/c	0.14	0.077	0.07	0.053	0.062
p (MPa)	8.05	9.71	9.86	9.86	9.55
$p/p_{0.2}$	1.034	0.75	0.769	0.8	0.775
σ_0 (MPa)	313	496	496	536	536
α	2.4	5.34	5.34	5.92	5.92
n	6.25	8.45	8.45	9.62	9.62
C	2.1	2.4	2.4	2	2.07
J_{cr} (N/mm)	487	432	432	439	439
$- T/\sigma_0$	0.672	0.575	0.544	0.606	0.611
$- Q$	0.667	0.591	0.546	0.648	0.651

α, Ramberg-Osgood exponent n, plastic constraint factor C, J-integral critical value J_{cr}, determined as J_m (corresponding to attaining the maximum force at the "force – force point displacement" curve), T-stress to yield stress ratio T/σ_0, and the Q parameter. Values of σ_0, α and n were derived from tensile tests, and the values of J_{cr} were derived from fracture tests run on CT specimens. Fracture pressure values p were read at the moment the ligament under the crack in the pipe body ruptured. Values for the plastic constraint factor on yielding, C, were determined on the basis of the J-integral in such a way that agreement was reached between the predicted and experimentally established fracture parameters for the given crack and fracture toughness of the material. The J-integral value was calculated using the GS method (Gajdoš&Srnec, 1994), on the one hand, and on the basis of the French nuclear code (RCC-MR, 1985), on the other.

It should be noted that in determining the C factor, the critical J-integral value established on CT specimens was considered – namely $J_{cr} = 439$ N/mm for steel X70, $J_{cr} = 432$ N/mm for steel X65 and $J_{cr} = 487$ N/mm for steel X52. It was found by a computational analysis of the CT specimens, employed to construct the R curve, that the Q parameter for these specimens was Q = 0.267. A comparison of this with the Q parameter for pipe bodies (Q ≈ -0.55 ÷ -0.65) reveals that the constraint in the CT specimens was much higher. This implies that the real fracture toughness –

i.e. the critical value of the J-integral, J_{cr} – was higher in the pipe bodies. The real C factor for a cracked pipe body is lower, so that the J-a curve for a pipe body is steeper than the curve for CT specimens with a greater C factor (Gajdoš&Šperl, 2011). Due to this, the J-integral for the axial part-through crack reaches the corresponding higher fracture toughness (for a lower constraint) for the same crack depth as the J-integral with a higher C factor reaches lower fracture toughness (determined on CT specimens). The situation is illustrated in Fig.3.

The normalized T–stress values in Table 1 were obtained using the plane solution – i.e. a solution for a crack of infinite length oriented longitudinally along the pipe. The problem was solved at the Institute of Physics of Materials, Brno, by the finite element method. The solution consisted of two steps: (i) a corresponding FEM network was established and corresponding boundary conditions were formulated for each crack depth, (ii) the magnitudes of the stress intensity factor and the T-stress were calculated for each FEM network by means of the CRACK2D FEM system with hybrid crack elements. The Q parameter values were derived from the $Q - T/\sigma_0$ curves obtained by O′Dowd and Shih (O′Dowd & Shih, 1991), by modified boundary layer analysis for different values of the strain coefficient (Ramberg-Osgood exponent, n). Strictly speaking, the Q parameter values from Table 1 do not correspond accurately to the values for the examined cracks, because the T-stresses were not computed for real semi-elliptical cracks, but for cracks spreading along the entire length of the pipe body ($a/c \approx 0$). Nevertheless, due to the fact that the ratio of the depth to the surface half-length of the examined cracks (a/c) was close to zero ($a/c=0.053 \div 0.14$), we can assume that the differences between the real values of the Q parameter and the values listed in Table 1 will be small.

The figures shown in Table 1 provide an idea of the nature of the changes both in the plastic constraint factor, C, and in the Q parameter brought about by changes in the relative crack length, a/t. The diagrams shown in Figs. 4 and 5 can be obtained on the basis of the graphic representation of the pairs $C - a/t$ and $Q - a/t$.

These diagrams clearly show the trends of the changes in the two parameters with a change in the relative crack depth, a/t. It follows that, in the range of relative depths examined here ($a/t = 0.57$ to 0.72), the plastic constraint factor, C, and the Q parameter are a growing function of the relative crack depth, a/t, the $Q - a/t$ dependence being rather weak. Expressed simply (i.e. linearly), the following relations are involved:

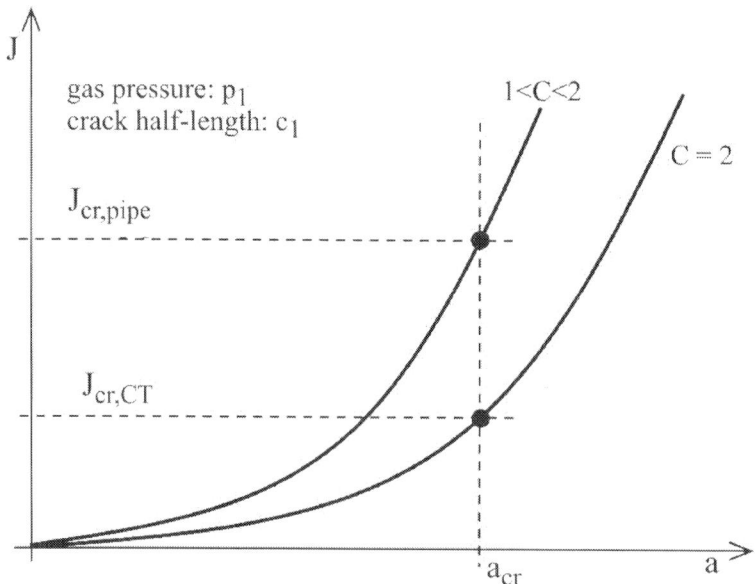

Figure 3. Schematic J-a dependence, (i) for a CT specimen, and (ii) for a pipe with an axial part-through crack

$$C = 2.56a/t + 0.53 \tag{18}$$

$$Q = 0.32a/t - 0.83 \tag{19}$$

The high scatter of the C and Q values is (i) due to differences in the cross section dimensions of the DN800 and DN1000 pipes and (ii) due to different values of the strain exponent n in the Ramberg-Osgood relation, because the pipes were made of three different materials.

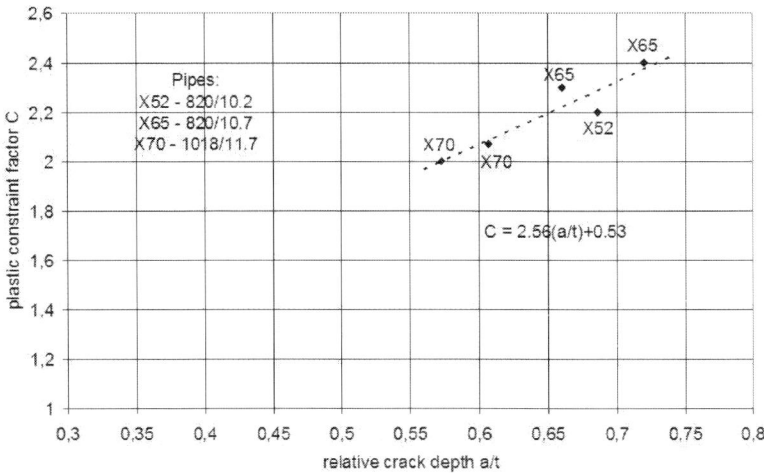

Figure 4. Plastic constraint factor, C, as affected by the relative crack depth, a/t

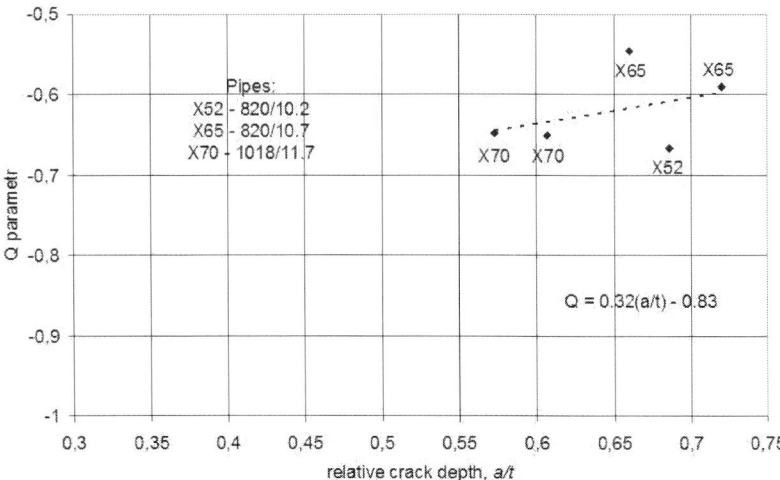

Figure 5. Parameter Q, as affected by the relative crack depth, a/t

Table 1 lists explicit values of Q and C for all examined cracks in the pipes that were used, and thus a graphic representation of the $C - Q$ relation can be plotted (Fig. 6). In the region where the established values of parameter Q for the examined pipe bodies are found, the $C - Q$ relation can be most simply described by the linear relation:

$$C = 3.4Q + 4.3 \qquad (20)$$

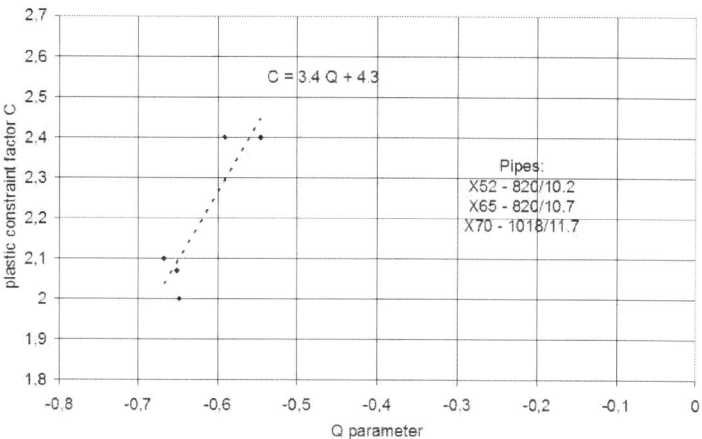

Figure 6. Dependence of the plastic constraint factor, C, on parameter Q

The relation implies that the plastic constraint factor, C, decreases with a decreasing value (increasing negative value) of parameter Q. The observed scatter of the experimental points is mainly due to inaccuracies of the T-stress estimate, which result from the substitution of the real conditions of cracks of certain lengths by the plane solution used in the task (crack along the entire length of the body).

FRACTURE TOUGHNESS

If we are to evaluate the strength reliability and the remaining life of gas pipelines, we need to get an accurate picture of the properties of the material that the gas pipelines are made of. In the case of gas pipelines operated for different periods of time, we should be aware that the properties of the m

In order to pass a qualified judgement on the reliability of a gas pipeline, we should know the true properties that the material displays at the time when the gas pipeline is being examined. The fracture properties can be characterized with sufficient generality by the fracture toughness, determined by quantities J_{in}, $J_{0.2}$ or J_m, where J_{in} is the so-called initiation magnitude of the J integral for a stable subcritical crack extension; $J0.2$ is the J magnitude corresponding to the real crack extension $\Delta a = 0.2$ mm, and J_m is the magnitude of the J integral corresponding to attaining the maximum force at the "force – force point displacement" curve. We should point here to two aspects of fracture toughness that can be encountered when dealing with pressure pipelines. One of them is the effect of pipe band straightening, and the other is the effect of stress corrosion cracks on fracture toughness.

The Effect of Straightening

Fracture toughness tests are carried out with fracture mechanical specimens, e.g. single edge notched bend (SENB) specimens or compact tension (CT) specimens. Both types are plane specimens. When investigating the integrity of thin-walled pressure pipelines, we face the problem of ensuring the planeness of the semiproducts for manufacturing the fracture mechanical specimens. The only way is press straightening of pipe bands taken from the pipe that is under investigation. As a consequence of the plastic deformation that the semi-product undergoes during straightening, internal stresses are induced not only in the semi-product but also in the final specimens. Therefore there are still some doubts about the reliability of the fracture toughness characteristics obtained with straightened specimens. In order to verify this matter, Gajdoš and Šperl (Gajdoš&Šperl, 2012) carried out an experimental investigation of fracture toughness, as determined using press straightened CT specimens and curved CT specimens, manufactured directly from a pipe band, i.e. ensuring that their natural curvature and wall thickness were preserved.

The so-called curved CT specimens (see Fig. 7) to some extent simulate the stress conditions in the pipe wall upon loading by internal pressure. In order to apply a circumferential force on these specimens, we used a special testing rig, similar to that developed by Evans (Evans et al., 1995). The rig is shown in Fig. 8.

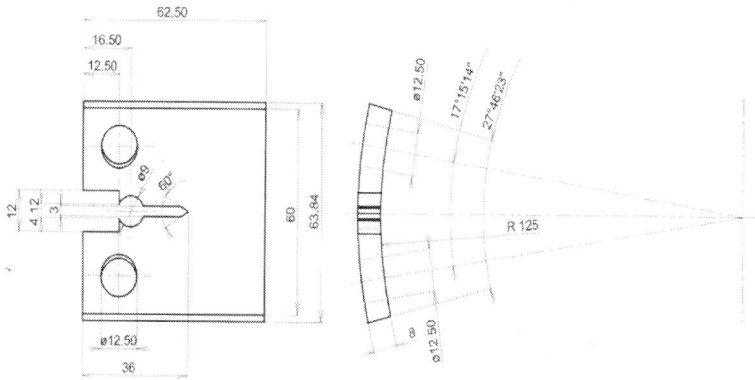

Figure 7. The shape and dimensions of the curved CT specimens

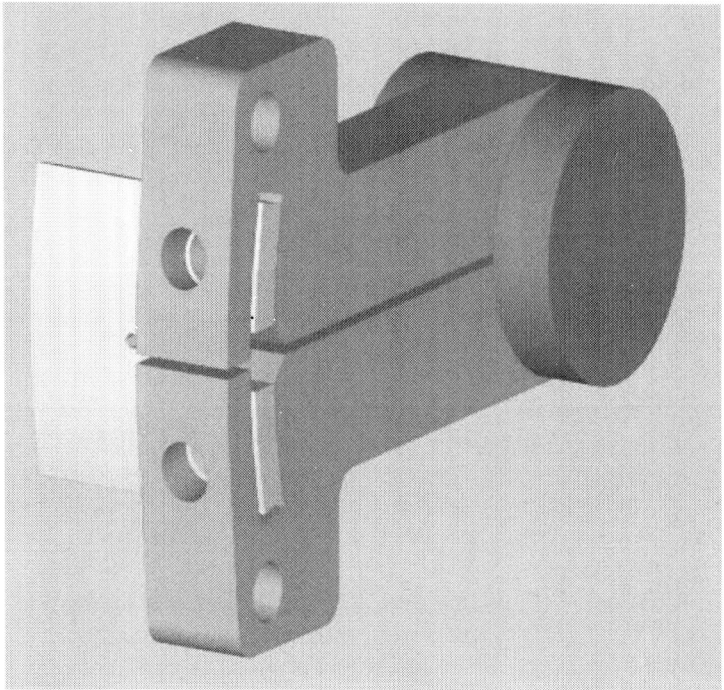

Figure 8. The testing rig for circumferential loading of a curved CT specimen

It is clear that the testing rig is tied with only certain cross – sectional dimensions of a pipe. In the case considered here, the dimensions corresponded to a pipe 266 mm in outside diameter and 8 mm in wall thickness. The material of the pipe was low-C steel CSN 411353. Static tests of the steel provided the following results: $R_{p0.2}$ = 286 MPa; R_m = 426 MPa; A_5 = 31%; Z = 54%. The Ramberg-Osgood constants had the following values: α = 6.23; n = 5.87; $\sigma 0$ = 286 MPa.

First, fracture toughness tests were carried out by an ordinary procedure, as specified in the ASTM standard (E 1820-01, 2001), on CT specimens manufactured from a press-straightened band taken from the pipe.

The result in the form of an R-curve is presented in Fig. 9. One point (designated by a triangle) has not been included in the regression analysis because it was outside the valid area of the diagram. The positions of J_{in} and $J0.2$ at the R-curve are clearly defined from the construction of the R-curve, the blunting line and the 0.2 offset line; the position of J_m is also indicated in the diagram, and it represents the mean of six values obtained on specimens where the maximum force was attained in loading the specimens. The R-curve determined by the least-square method is described by a power function (20):

$$J = 327.05 (\Delta a)^{0.6406} \qquad (21)$$

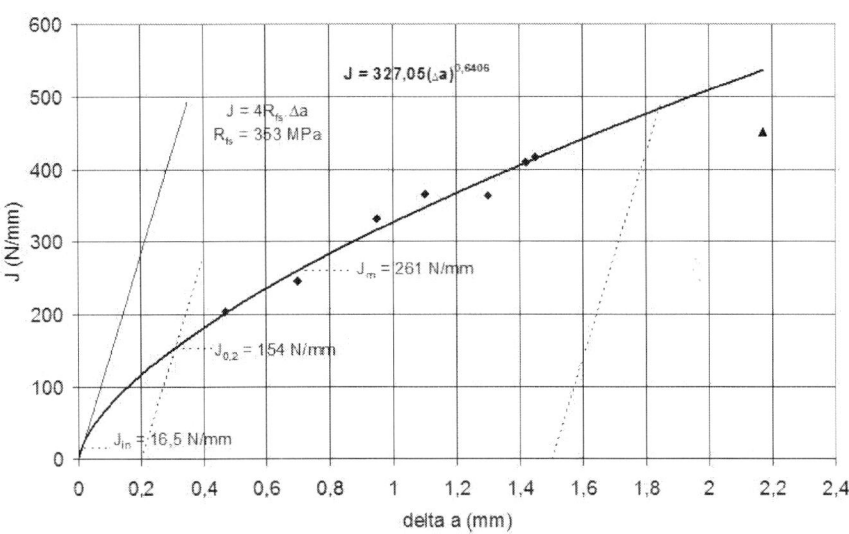

Figure 9. R curve for CT specimens manufactured from a press-straightened semi-product

Eight specimens were used for fracture-mechanical tests of curved CT specimens. Cracks were cycled up at a frequency of 4 Hz, using the testing rig, Fig. 8. The stress state in the inner side of a specimen was bigger than in the outer side, because of the bending moment induced by the out-of-axis action of the vertical component of the tangential force with regard to the intersection of the middle cylindrical area of the specimen with the symmetry plane of the specimen. For this reason, the growth rate of the fatigue crack was higher in the inner side than in the outer side.

This resulted in uneven length of the fatigue crack on the two sides of a specimen after finishing the cycling up. On one half of the specimens, a slant front of the starting notch was therefore made in such a way that the notch was 1 mm deeper on the outer side. By this operation, a much more even front of the fatigue crack was obtained. This is clearly demonstrated in Figs. 10 and 11, which show the fracture surfaces of specimens with a straight front and a slant front of the starting notch. In the two photographs, we can observe areas corresponding to the notch, fatigue, static crack extension and final break after the specimens were cooled down in liquid nitrogen.

On the basis of the finite element analysis and the compliance measurements made by Evans (Evans et al., 1995), it was concluded that the use of standard expressions for determining K factor will not cause error greater than 4% for curved CT specimens. By proceeding in the same way as in standard $J - \Delta a$ testing, an R-curve was obtained for curved CT specimens. It is described by a power function (21), and is presented in Fig. 12.

$$J = 278.21 (\Delta a)^{0.525} \tag{22}$$

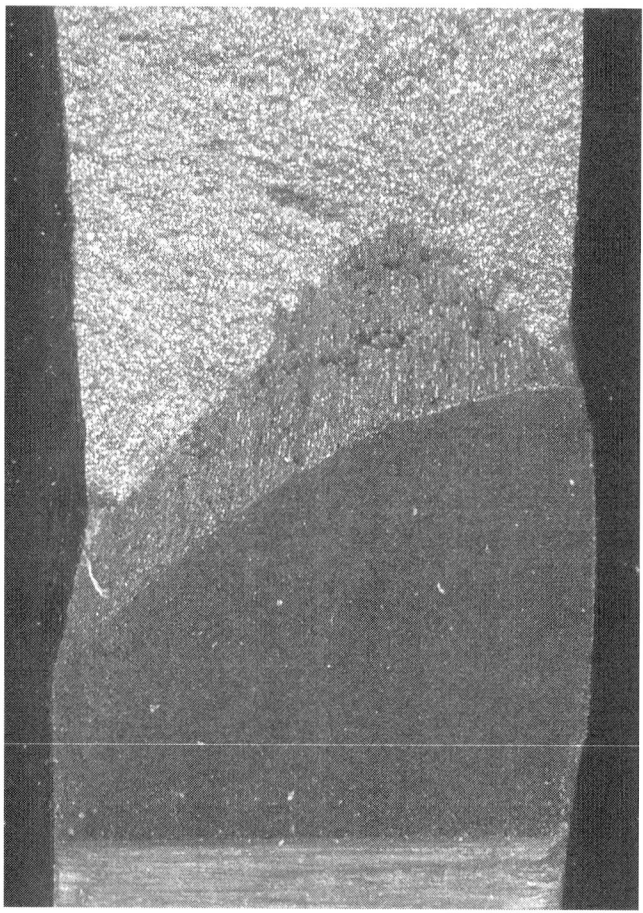

Figure 10. Fracture surface – straight front of the notch

Figure 11. Fracture surface – slant front of the notch

A comparison of the two R-curves shows that the decline of the R-curve obtained with the curved CT specimens is less than the decline of the R-curve obtained with plane, i.e. straightened, CT specimens. The higher decline of the R-curve with the straightened CT specimens is most probably connected with work hardening of a semiproduct during straightening. In the mathematical description of the R-curve of the curved CT specimens, not only the exponent but also the constant is less than for the standard R-curve. This means that the standard R-curve is situated above the R-curve of the curved CT specimens. However, the lower position of the R- curve for the curved CT specimens does not mean significantly lower magnitudes of the fracture toughness characteristics. For example, the J_m value is lower by 1.1%, the $J_{0.2}$ is lower by less than 3%, and the magnitude J_{in} is even higher than the respective characteristics for plane (straightened) CT specimens. In absolute units, the difference is 2.9 N/mm for J_m and 4.6 N/mm for $J_{0.2}$. There is a significant difference in J_{in}, namely 29.7 N/mm in favour of the curved CT specimens.

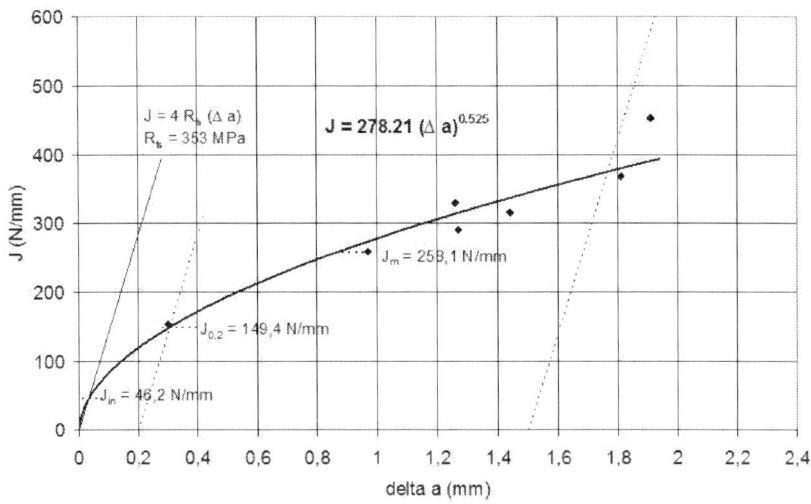

Figure 12. R-curve for curved CT specimens

By accounting the scatter of the results in the form of the J - Δa points, caused both by a natural process of subcritical crack growth and by inaccuracies in determining the J-integral and, in particular, the crack extension during monotonic loading of a specimen, it can be stated with a high level of reliability that the fracture toughness of a pipe material determined on straightened CT specimens is practically the same as the fracture toughness determined on curved CT specimens.

The Effect of Stress Corrosion

(Gajdoš et al., 2011) investigated the stress corrosion fracture toughness of gas pipeline material, and compared it with fatigue fracture toughness. The material used for the investigation was a low-C steel according to CSN 411353 (equivalent to ASTM A519), containing 0.17% C, 0.035% P, 0.035% S. The test CT specimens were manufactured from a real pipe section cut out from a DN 150 gas pipeline 4.5 mm in wall thickness while it was being repaired after 20 years of operation. Before the CT specimens were manufactured, the pipe section was press straightened. Owing to the small thickness of the specimens (a low constraint), the fracture toughness values cannot be qualified to represent the real fracture toughness values. However, they can be used as a comparative measure of fracture toughness, thus enabling quantification of the effect of stress corrosion cracks on the apparent fracture toughness.

The CT specimens were first cyclically loaded by a routine procedure used in determining fracture toughness; the only difference was that the cycling was stopped when the growth of the fatigue crack reached approximately the magnitude $\Delta a_{FA} \approx 1.5$ mm. After that, the CT specimens were put into the stress-corrosion (SC) crack generator with an acidic solution according to the NACE Standard (NACE Standard TM0177, 2005). This solution consisted of 50 g NaCl (sodium chloride) + 5 g CH_3COOH (acetic acid) + 945 g H_2O, and during the generating process it was bubbled by H_2S (hydrogen sulphide). A constant force F of 3 kN was applied to the specimens. The corresponding level of the nominal stress (tension and bending) at the fatigue crack tip exceeded the yield stress $R_{p0.2}$ by about 25%. The crack length increment due to stress-corrosion Δa_{SC} was determined with the help of the relations for elastic crack-edge displacements at CT specimens. In total, three groups of CT specimens were prepared. The first group (A) was the reference group; the specimens from this group contained only the fatigue crack. The second group of CT specimens (B) contained specimens that were left freely in air at the indoor temperature for two weeks after being removed from the SC crack generator, and were then subjected to fracture toughness tests. The specimens from the third group (C) were tested immediately after they had been removed from the SC crack generator (the time difference between testing the first specimen and the last specimen being approximately 20 minutes).

The results confirmed that the fracture resistance of a component (given by the apparent fracture toughness) depends not only on the material of the component and on the crack tip constraint (the thickness of the wall of the component) but also on the origin of the crack (fatigue, stress corrosion), and thus on the corresponding crack growth mechanism. In contradiction with the opinion that low-C steels are not susceptible to stress corrosion cracking our results showed that under conditions specified in (NACE Standard TM0177, 2005) stress corrosion cracks can also be generated from fatigue cracks in low-C steels such as CSN 411353. Unlike a fatigue crack, the occurrence of a stress-corrosion crack in a component means a significant decrease in the fracture toughness characteristics while the crack is exposed to stress corrosion conditions, and a partial "recovery" of the fracture toughness when the stress corrosion conditions are removed. The results for all three groups of specimens are summarized in Fig. 13.

Figure 13. A bar chart of the J integral values for specimens of groups A, B and C

As this figure shows, the stress corrosion fracture toughness characteristics for the low-C steel CSN 411353 were lower than the fatigue fracture toughness characteristics by a factor ranging between 4.5 (J_m value) and 5.7 (J_{in} value). However, a two-week recovery period made it possible to recover their fracture properties to some extent, namely the J-integral J_m to almost 80%, the J-integral $J_{0.2}$ to about 60%, and the J-integral J_{in} to about 22% of the fatigue crack J-integral values. It follows from here that in evaluating the reliability of gas pipelines it is always necessary to examine the character of the cracks in the pipe wall, and in the case of stress corrosion cracks to take into account that the fracture toughness can be drastically lower than the values determined on specimens with cracks of fatigue origin.

BURST TESTS

An experimental verification of the fracture conditions of gas pipelines can be made most accurately on a test pipe body cut out of the gas pipeline to be examined. When deciding on the length of the test pipe body, we should bear in mind that the working length of the body (characterized by the absence of stress effects from welded-on bottoms) will be shorter by $2 \times 2.5 \sqrt{(Rt)} \approx 3.5 \sqrt{(Dt)}$. It is usually sufficient for the distance between the welds of dished bottoms to be at least 3.5D. This length permits a number of starting cuts to be placed axially along the length of the body. The cuts are made to initiate crack growth when the body is subsequently pressurized by a fluctuating pressure. The cuts can be made in several ways, one of which uses a thin grinding wheel. The smallest real functional thickness of such a wheel is about 1.2 mm, and the corresponding width of the cuts made with it is approximately 1.5 mm. Depending on the type of pipes of which gas pipelines are built (seamless, spirally welded, longitudinally welded), the starting cuts can be provided in the base material, in the transition region or in the weld metal, their orientation being axial, circumferential or along the spiral weld.

Preparation of Test Pipe Bodies

It is appropriate to relate the surface length of the cuts to the wall thickness of the pipe body. Testing the body for the danger posed by so-called long cracks should be carried out with crack lengths not exceeding twenty times the wall thickness of the pipe body. The situation with the depth of the starting cuts is different. The depth of an initiated fatigue crack must be at least 0.5 mm along the whole perimeter of the cut tip, so that the cut with the initiated crack at its tip can be considered as a crack after the pipe body has been subjected to cycling. This value follows from the work done by Smith and Miller (Smith & Miller, 1977). If such a crack a_t in size finds itself in a notch root defined by depth a_v and radius of the roundness ρ (see Fig. 14), this configuration can be regarded as a surface crack a_e in depth, where

$$a_e = \left(1 + 7.69\sqrt{\frac{a_v}{\rho}}\right) a_t \quad for \quad a_t < 0.13\sqrt{a_v\rho}$$

$$a_e = a_v + a_t \quad for \quad a_t \geq 0.13\sqrt{a_v\rho}$$

$$(23)$$

It is evident that for $a_t \geq 0.13\sqrt{a_v\rho}$, a cut with a crack along the perimeter of the cut tip can be taken for a crack with a depth of $a_v + a_t$. For the cut width $2\rho = 1.5$–2.0 mm and the notch depth $a_v = 6$–10 mm (in relation to the wall thickness), we find that the fatigue increment of the size of the initiated crack, a_t, should be greater than about 0.5 mm.

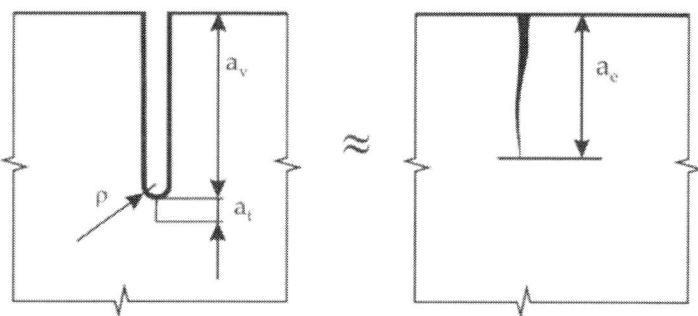

Figure 14. Substitution of a notch with a crack by the equivalent crack

As described in paragraph 3.2, three test pipe bodies, made of X52, X65 and X70 steels, were provided with working slits and so-called check slits, which were of the same surface length as the working slits but their depth was greater. These check slits functioned as a safety measure to prevent cracks that developed at the working slits from penetrating through the pipe wall. For illustration, a DN1000 test pipe body with a working length of 3.5 m is shown in Fig. 15. The check slits are denoted in Fig. 15 by a supplementary letter K. The material of the test pipe body is a thermo-mechanically treated steel X70 according to API specification. The pipe is spirally welded, the weld being inclined at an angle of $\varphi = 62°$ to the pipe axis. It is provided with starting cuts oriented either axially or in the direction of the strip axis (i.e. in the direction of the spiral) and then along or inside the spiral weld. The cuts differ in length ($2c = 115$ mm or 230mm) and in depth ($a = 5, 6.5, 7,$ and 7.5 mm). We are particularly interested in axial (longitudinal) slits situated aside welds, because these are sites where axial cracks will be formed in the basic material of the pipe.

Efforts were made in the fracture tests to keep the circumferential fracture stress below the yield stress, because the operating stress in gas pipelines is virtually around one half of the yield stress (and does not exceed two-thirds of the yield stress even in intrastate high-pressure gas transmission pipelines). Calculations reveal that in order to comply with this, the depth of the axial semi-elliptical cracks should be greater than one half of the wall thickness. Oblique cracks should be even deeper, as the normal stress component opening these cracks is smaller. If the crack depth is to have a certain magnitude before the fracture test is begun, the depth of the starting slit should be smaller than this magnitude by the fatigue extension of the crack along the perimeter of the slit tip. At the same time, we should bear in mind that the greater the fatigue extension of the crack, the better the agreement with a real crack.

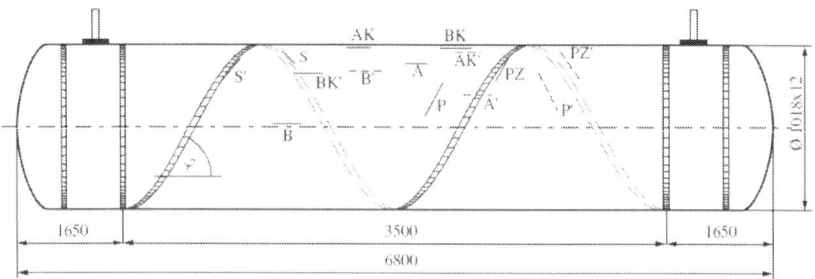

Figure 15. Test pipe body with the starting cuts marked

PREDICTION OF FRACTURE PARAMETERS

After the starting slits were made, the test pipes were subjected to water pressure cycling to produce fatigue cracks in the tips of the starting slits. The cycling was carried out in a pressurizing system, which included a high-pressure water pump, a collecting tank, a regulator designed to control the amount of water that was supplied and, consequently, the rate at which the pressure is increased in the pipe section. This was effected by opening by-pass valves.

In cycling the cracks, the water pressure fluctuated between $p_{min} = 1.5$ MPa and $p_{max} = 5.3$ MPa, and the number of pressure cycles was between 3 000 and 4 000. The period of a cycle was approximately 150 seconds. The cycling continued until a crack initiated in one of the check slits became a through crack. This moment was easy to detect, because it was accompanied by a water leak. By choosing an appropriate difference between the depths of the working slits and the check slits it was possible to obtain a working crack depth (= starting slit depth + fatigue crack extension) of approximately the required size. To run a test for a fracture, however, it was necessary to remove the check slit which had penetrated through the wall of the test pipe from the body shell and to repair the shell, e.g. by welding a patch in it.

After removing the check slit with a crack which penetrated through the wall, and repairing the shell of the test pipe, the pipe was loaded by increasing the water pressure to burst. The test procedure, which was common for all test pipes, will now be briefly described for the DN1000 pipe shown in Fig. 15. As the figure suggests, slits A, A′, B and B′ were oriented along the axis of the pipe. The nominal length of notches B, B′ was twice as long as notches A, A′, but notches B, B′ were shallower. As was mentioned above, the cracks at the slit tips were extended by fluctuating water pressure, and this proceeded until the cracks from the check slits (BK, BK′) grew through the wall and a water leak developed. Then the damaged parts of the shell were cut out, patches were welded in their place, and the test pipe was monotonically loaded to fracture at the location of crack B or B′. The burst of the test pipe at crack B is shown in Figs. 16 and 17 (as a detail). A part of the fracture surface is shown in Fig. 18.

Figure 16. Burst initiated on slit B with a fatigue crack

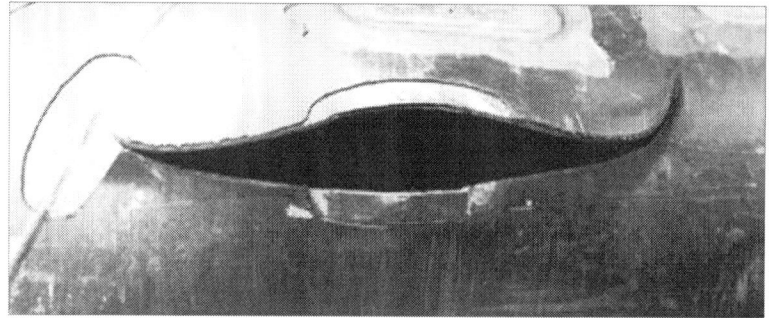

Figure 17. Burst initiated on slit B – a detail

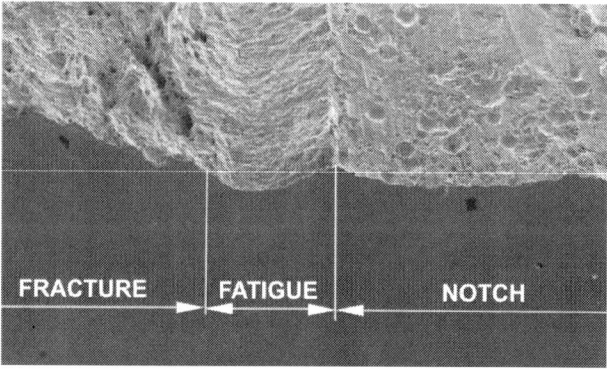

Figure 18. A part of the fracture surface of crack B (fatigue region ~ 2.4 mm)

Evidently, at the instant of fracture the crack spread not only through the remaining ligament, but also lengthwise. After removing the part of the pipe shell with crack B, a patch was welded in and the second burst test followed. Table 2 extracts from Table 1 the numerical values of the geometrical parameters, the J-integral fracture values, the Ramberg-Osgood constants, the fracture pressure and the fracture depth for cracks B and B′, respectively.

Table 2. Some characteristics referring to crack B and crack B′

Characteristics	Crack B	Crack B′
CRACK DIMENSIONS		
half-length, c (mm)	115	127
depth in fracture, a_f (mm)	7.1	6.7
RAMBERG-OSGOOD PARAMETERS		
$\alpha / n / \sigma_0$ (MPa)	5.92 / 9.62 /536	5.92 / 9.62 /536
FRACTURE TOUGHNESS		
$J_{cr} = J_m$ (N/mm)	439	439
FRACTURE PRESSURE		
p_f (MPa)	9.55	9.86

It should be noted that Table 2 includes the Ramberg-Osgood constants for the circumferential direction of the test pipe, with the crack oriented axially in the pipe. This is because the stress-strain properties perpendicular to the crack plane are crucial in determining the J-integral for an axial crack. The stress-strain dependence in the circumferential direction should therefore be taken into account where an axial orientation of the crack is concerned. The most important fracture test results from the viewpoint of the fracture conditions are the magnitudes of the fracture pressure, pf, and the fracture depth, af, for a given crack length 2c. It follows from Table 2 that $p_f = 9.55$ MPa and $a_f = 7.1$ mm for crack B, and $p_f = 9.86$ MPa and $a_f = 6.7$ mm for crack B′. These values are also shown in the last two columns of Table 1.

Now let us predict the fracture conditions according to engineering approaches, and compare the prediction results with the real fracture parameter values (pressure, crack depth). The procedure for verifying the engineering methods for the predictions involves determining either the fracture stress for a given (fracture) crack depth, or the fracture crack depth for a given (fracture) pressure. To illustrate this, we select the latter case – i.e. determining the fracture depth of a crack for a given (fracture) pressure. Fig. 19 shows the J-integral vs. crack B depth dependences, as determined by the FC and GS predictions for the fracture hoop stress given by the measured fracture pressure. When using equations (9), (10), and (12) to determine J-integrals, the following parameters were used for the calculation: D = 1018 mm; t = 11.7 mm ; p = p_f = 9.55 MPa; c = 115 mm; α = 5.92; n = 9.62; σ_0 = 2.07×536 = 1110 MPa (i.e. C = 2.07). Fig. 20 shows similar dependences for crack B′.

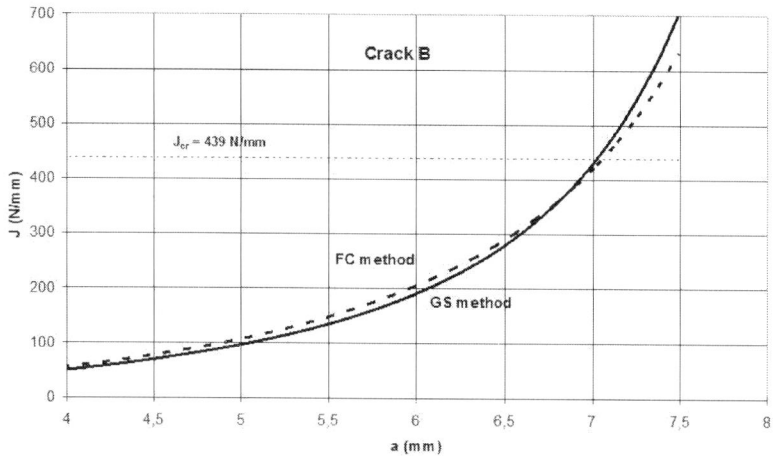

Figure 19. Prediction of the fracture depth for crack B (p = p_f = 9.55 MPa and C = 2.07)

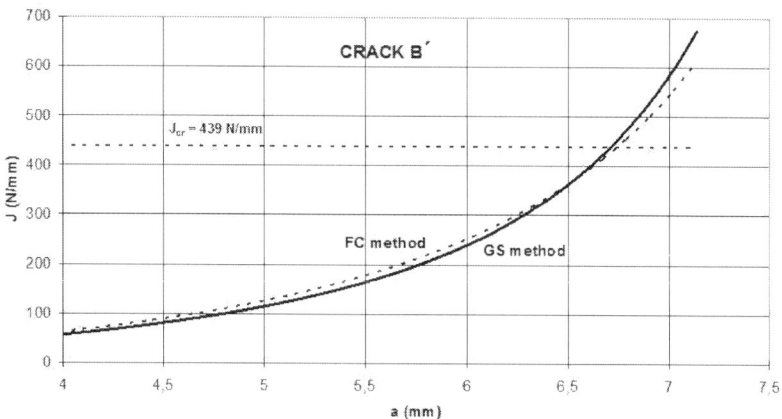

Figure 20. Prediction of the fracture depth for crack B′ (p = p_f = 9.86 MPa and C = 2.0)

The same computational parameters as those employed in the case of crack B were used in the equations to determine the J-integral according to the FC and GS methods, with the exception of the fracture pressure ($pf = 9.86$ MPa), the crack half-length (c = 127 mm) and factor C (C = 2.0). As is evident from Fig. 19, the intersection of the straight line J = J_{cr} = 439 N/mm with the two J − a curves gives the value $a_{cr} \approx 7.05$ mm, which is well consistent with crack depth B a_{cr} = 7.1 mm, established experimentally. Similarly, the intersection of the straight line J = J_{cr} = 439 N/mm with the J−a curves according to the FC and GS procedures in Fig. 20 shows the fracture crack depth a_{cr} to be virtually identical to the experimentally found fracture depth a_f = 6.7 mm. For other test pipes, namely DIA 820/10.7, made of X65 steel, and DIA 820/10.2, made of X52 steel, various magnitudes of the plastic constraint factor C were obtained to achieve good agreement of the geometric parameters at fracture with the experimental parameters. They are illustrated in Fig. 4. The conclusion can thus be drawn that very good agreement of the fracture parameter values predicted by the FC and GS engineering approaches with the values found experimentally can be achieved when using the plastic constraint factor on yielding, C, at the level C = 2. If a higher value of the C factor provides more precise results, the use of the value C = 2 will yield a conservative result.

CONCLUSION

A specific fracture-mechanics-based procedure for assessing the integrity of pressurized thin-walled cylindrical shells made from steels includes a theoretical treatment for cracks in pipes. On the basis of both experimental work and a fracture-mechanical evaluation of experimental results, an engineering method has been worked out for assessing the geometrical parameters of critical axial crack-like defects in a high-pressure gas pipeline wall for a given internal pressure of a gas. The method makes use of simple approximate expressions for determining fracture parameters K, J, and it accommodates the crack tip constraint effects by means of the so-called plastic constraint factor on yielding. Involving this in the fracture analysis leads to multiplication of the uniaxial yield stress by this factor in the expression for determining the J-integral. Two independent approximate equations for determining the J-integral provided very close assessments of the critical geometrical dimensions of part-through axial cracks. With the use of the crack assessment method, the critical gas pressure in a pipeline can also be determined for a given crack geometry.

The fracture toughness with which the J-integral is compared in fracture analysis is determined using fracture mechanics specimens (e.g. CT,

SENB and others). Experiments made on press-straightened CT specimens and on curved CT specimens with a natural curvature, made from pipe 266/8 mm of low-C steel CSN 411353, showed that straightening a pipe band prior to the machining of CT specimens had a practically negligible effect on the fracture toughness characteristics ($J_{0.2}$, J_m). However, experiments with fracture toughness testing of specimens with stress corrosion cracks, formed by the hydrogen mechanism, showed a dramatic reduction of all fracture toughness characteristics in comparison with fracture toughness determined on specimens with fatigue cracks, e.g. the quantities J_{in}, $J_{0.2}$ and J_m dropped to 17.5%, 18.5%, and 22.3%, respectively. A partial "recovery" of fracture toughness characteristics was observed when the stress corrosion conditions were removed.

ACKNOWLEDGEMENT

Financial support from Research Plan AV0Z 20710524 and from grant-funded projects GACR P105/10/2052 and P105/10/P555 are highly appreciated.

REFERENCES

1. AinsworthR. A. & O´Dowd, N. P. (1995Constraint in the Failure Assessment Diagram Approach for Fracture Assessment. Transactions ASME- Journal of Pressure Vessel Technology, 117260267
2. ASTM Standard E 1820-01 (2001Standard Test Method for Measurement of Fracture Toughness, 2001
3. F Erdogan, J. J Kibler, 1969Cylindrical and Spherical Shells with Cracks. International Journal of Fracture Mechanics, 53229237
4. F Erdogan, F Delale, J. AOwczarek, 1977Crack Propagation and Arrest in Pressurized Containers. Journal of Pressure Vessel Technology, 99February 1977), 9099
5. J. T Evans, G Kotsikos, R. F Robey, 1995A method for Fracture Toughness Testing Cylinder Material. Engineering Fracture Mechanics, 502295300
6. E. S Folias, 1969On the Effect of Initial Curvature on Cracked Flat Sheets. International Journal of Fracture Mechanics, 54327346
7. E. S Folias, 1970On the Theory of Fracture of Curved Sheets. Engineering Fracture Mechanics, 22151164
8. ĽGajdoš, M Srnec, 1994An Approximate Method for J Integral Determination. ActaTechnica CSAV, 392151171

9. GajdošĽubomíret al2004Structural Integrity of Pressure Pipelines. Transgas, 80-86616-03-7, Prague, Czech Republic

10. ĽGajdoš, M Šperl, 2011Application of a Fracture-Mechanics Approach to Gas Pipelines. Proceedings of World Academy of Science, Engineering and Technology. 73January 2011, 480487

11. ĽGajdoš, M Šperl, J Siegl, 2011Apparent Fracture Toughness of Low-Carbon Steel CSN 411353 as Related to Stress Corrosion Cracks. Materials and Design, 328-943484353

12. ĽGajdoš, M Šperl, 2012The Effect of Straightening a Curved Body on Its Fracture Properties (in Czech). In: Proceedings of the 21st Colloquium "Safety and Reliability of Gas Pipelines", Prague, 2012

13. J. W Hutchinson, 1968Singular Behaviour at the End of a Tensile Crack Tip in a Hardening Material. Journal of the Mechanics and Physics of Solids, 161331

14. I Milne, R. A Ainsworth, A. R Dowling, A. T Stewart, 1986Assessment of the Integrity of Structures Containing Defects. CEGB Report No. R/H/R6Rev.3, Central Electricity Generating Board, London, U.K., (1986)

15. NACE Standard TM0177 (2005Laboratory Testing of Metals for Resistance to Sulfide Stress Cracking and Stress Corrosion Cracking in H2S Environments. Item 21212

16. J. C Newman, 1973Fracture Analysis of Surface and Through-Cracked Sheets and Plates. Engineering Fracture Mechanics, 53667689

17. O Dowd, N. P Shih, C. F. (1991Family of Crack-Tip Fields Characterized by a Triaxiality Parameter- I. Structure of Fields. Journal of the Mechanics and Physics of Solids, 398981015

18. RCC- MR (1985Design and Construction Rules for Mechanical Components of FBR Nuclear Island. First Edition (AFCEN-35Av. De Friedeland Paris 8), (1985)

19. J. R Rice, G. F Rosengren, 1968Plane Strain Deformation near a Crack Tip in a Power-Law Hardening Material. Journal of the Mechanics and Physics of Solids, 16112

20. P. M Scott, T. W Thorpe, 1981A Critical Review of Crack Tip Stress Intensity Factors for Semi- Elliptical Cracks. Fatigue of Engineering Materials and Structures, 44291309

21. R. C Shah, A. S Kobayashi, 1973Stress Intensity Factors for an Elliptical Crack Approaching the Surface of a Semi- Infinite Solid. International Journal of Fracture, 9133146

22. C. F Shih, O Dowd, N. P Kirk, M. T. (1993A Framework for Quantifying Crack Tip Constraint. In: Constraint Effects in Fracture. ASTM STP 1171, American Society for Testing and Materials, Philadelphia, 220

23. R. A Smith, K. J Miller, 1977Fatigue Cracks at Notches. International Journal of Mechanical Sciences, 191122

CITATION

L'ubomírGajdoš and Martin Šperl (2012). Evaluating the Integrity of Pressure Pipelines by Fracture Mechanics, Applied Fracture Mechanics, Dr. Alexander Belov (Ed.), ISBN: 978-953-51-0897-9, InTech, DOI: 10.5772/51804.

CHAPTER 7

Multi-Physics Dynamics of a Mechanical Oscillator Coupled to an Electro-Magnetic Circuit

Ioannis T. Georgiou[1, 2], Francesco Romeo[3]

[1] School of Naval Architecture and Marine Engineering, National Technical University of Athens, Greece
[2] Adjunct Faculty Member, School of Mechanical Engineering, Purdue University, West Lafayette, IN, United States
[3] Department of Structural and Geotechnical Engineering, SAPIENZA University of Rome, Via Gramsci 53, 197 Rome, Italy

ABSTRACT

The dynamics of a non-linear electro-magneto-mechanical coupled system is addressed. The non-linear behavior arises from the involved coupling quadratic non-linearities and it is explored by relying on both analytical and numerical tools. When the linear frequency of the circuit is larger than that of the mechanical oscillator, the dynamics exhibits slow and fast time scales. Therefore the mechanical oscillator forced (actuated) via harmonic voltage excitation of the electric circuit is analyzed; when the forcing frequency is close to that of the mechanical oscillator, the long term damped dynamics evolves in a purely slow timescale with no interaction with the fast time scale. We show this by assuming the existence of a slow invariant manifold (SIM), computing it analytically, and verifying its existence via numerical experiments on both full- and reduced-order systems. In specific regions of the space of forcing parameters, the SIM is a complicated geometric object as it undergoes folding giving rise to hysteresis mechanisms which create a pronounced non-linear resonance phenomenon. Eventually, the roles played by the electro-magnetic and mechanical components in the resulting complex response, encompassing bifurcations as well as possible transitions from regular to chaotic motion, are highlighted by means of Poincaré sections.

INTRODUCTION

This paper addresses the dynamics of a system in which a mechanical linear oscillator is non-linearly coupled to a linear electric circuit through an electromagnet. The interest on this type of systems stems from their growing use in mechatronics for a variety of applications such as signal processing, sensing and actuation [1]. Depending on the field of application, mechanical vibratory devices integrated with electronics are made of components ranging in size from nano to micro and macro dimensional scales. Examples of such systems are abundant in modern engineering design, ranging from electric propulsion of land vehicles and ships to micro electromechanical and biomedical devices [2] and [3]. Regardless of their scale they represent an interesting class of dynamical systems as they involve multi-physics dynamics encompassing coupling among the mechanical, electrostatic, electromagnetic and fluid fields [4], [5] and [6]. Lumped- or distributed-parameters analytical models, finite-element models and reduced-order models have been proposed for modeling and simulate these multi-field dynamical systems. Moreover, as it is well known, these devices are prone to exhibit non-linear behavior arising either from their inherent mechanical nature and/or from the coupling itself. For small scale devices (M/NEMS), the mechanical subsystem design includes beams, plates or lumped masses that are often characterized by low damping, high flexibility and the presence of non-linear potential fields; thus, the resulting dynamics is strongly influenced by unavoidable non-linear behaviors that can be even actively exploited. An exhaustive overview of research activity on non-linear behaviors arising in micro- nano/resonators can be found in [7]. In this review a number of studies addressing directly and parametrically excited resonators as well as arrays of coupled M/NEMS exhibiting synchronization and vibration localization are reported. A more recent comprehensive overview of MEMS dynamics can be found in [8], in which the sources of non-linearities arising in MEMS modeling due to forcing, damping and stiffness are discussed.

As far as the coupling between electrical and mechanical subsystems, the non-linear nature of the electro-magnetic coupling force has to be taken into account. In this respect, an asymptotic approach for self-excited oscillations was discussed in [9] for a system modeled through two coupled second order differential equations. More recently, in [10], the linear and weakly non-linear dynamics of a coupled electro-mechanical system was studied. In this paper the mechanical system was modeled by a standard second-order differential equation with respect to displacement whereas the electro-magnetic one, after neglecting the capacitance term, was governed by a first-order differential equation with respect to current.

The near-resonant vibrations resulting from forcing the mechanical oscillator were analyzed by means of a classical perturbation method. Softening behavior and jump phenomena were thereby parametrically investigated with respect to the electro-mechanical and the electro-magnetic coupling parameters.

The system analyzed in this paper is modeled following [9] and the dynamics resulting from exciting the mechanical oscillator via harmonic voltage applied to the electric circuit is considered. The global dynamics of such multi-physics system can be effectively handled to make it functioning either as a sensor or an actuator for applications in the micro electromechanical context. After having introduced the governing differential problem in Section 2, the slow–fast time decomposition analysis is addressed in Section 3. Next, in Section 4, the interaction between the slow and fast dynamics is numerically explored leading, in Section 5, to the analytic computation of the Slow Invariant Manifold (SIM). The main features of the slow dynamics of the full order system (FOS) are described in Section 6. In the following section the approximations to the SIM are described up to the tenth order reduced slow system and the comparison between the various order reduced systems and the full one is carried out by means of bifurcation diagrams and frequency–amplitude plots. The complex chaotic interaction dynamics is eventually addressed in Section 8 through Poincaré sections and concluding remarks emerging from the analysis are reported.

A MULTI-PHYSICS COUPLED OSCILLATORS SYSTEM

In Fig. 1 a dynamical system consisting of coupled mechanical and electro-magnetic subsystems is sketched. The mechanical part is a linear oscillator coupled non-linearly through an electromagnet to a linear electric circuit. Let m, c, k denote respectively the mass, dissipation, spring parameters of the linear oscillator. Let L, C, R denote respectively the inductance, capacitance, and resistance of the electrical oscillator. The variables x and q denote the mechanical and electrical displacement (charge), respectively.

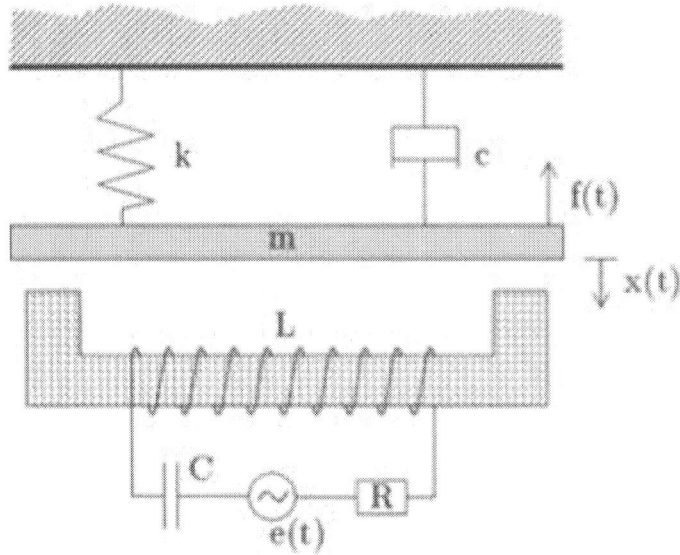

Figure 1. Mechanical system coupled via a magnetic field to a magneto-electrical system.

The velocity of the mass and the electric current are $v = \dot{x}, i = \dot{q}$, respectively. The variables f and e denote external mechanical forcing and voltage excitation, respectively. This multi-physics coupled system is a non-linear dynamical system with its coupling non-linearity stemming from the dependence of the inductance on the displacement and possibly velocity of the metallic oscillator mass. In general, as known from electro-magnetic theory, the inductance is a function of the displacement of the metallic mass. In this work we assume that the magnet is positioned at a characteristic distance l_0 from the equilibrium position of the oscillating metallic mass. We assume that the inductance is approximated by a linear function of the metallic mass displacement [9]:

$$L(x) = L_0 + L_1 x \tag{1}$$

in which $0 < x < l_0$. The linear natural frequencies and linear dissipation factors of the mechanical and electrical oscillators are given, respectively, by

$$\omega_m^2 = \frac{k}{m}, \quad \omega_e^2 = \frac{1}{L_0 C}, \quad \zeta_m = \frac{c}{2\sqrt{km}}, \quad \zeta_e = \frac{R\sqrt{C}}{2\sqrt{L_0}} \tag{2}$$

The electro-mechanical inertia and non-linear inductance coupling parameters are defined as

$$\epsilon = \frac{L_0}{2m}, \quad \alpha = \frac{L_1}{L_0} \tag{3}$$

We introduce the external force-per-mass and voltage-per-inductance excitations:

$$\hat{f} = \frac{f}{m}, \quad \hat{e} = \frac{e}{L_0} \tag{4}$$

The equations of motion of the coupled electro-magneto-mechanical system are given by the following two coupled second order ordinary differential equations:

$$\ddot{x} + 2\zeta_m \omega_m \dot{x} + \omega_m^2 x = \epsilon \alpha \dot{q}^2 + \hat{f}$$
$$(1 + \alpha x)\ddot{q} + (2\zeta_e \omega_e + \alpha \dot{x})\dot{q} + \omega_e^2 q = \hat{e} \tag{5}$$

These equations are similar to those in reference [9] where self-excited oscillation of an analogous system is addressed. The dynamics of the system excited by harmonic voltage while the external forcing in the mechanical part is absent, i.e. $\hat{f} = 0$, is studied. The interest lies in investigating how the mechanical part will respond when it is actuated by the electrical part; the question being to ascertain whether the steady dynamics of the former is determined by the dynamics of the latter. The study is carried out by combining direct numerical simulations and geometric mechanics formulations in the form of invariant manifolds of the system response. Invariant manifolds furnish global picture of the dynamics involving the time scales of the mechanical and the magneto-electrical subsystems. The slow–fast decomposition approach presents a global geometric picture of the dynamics and it allows anticipating bifurcations due to interactions between the oscillators.

SLOW–FAST TIME DECOMPOSITION ANALYSIS

The theory of Geometric Singular Perturbations (GSP) [11] provides the proper mathematical framework to support geometrically a slow–fast decomposition of the coupled multi-physics dynamics. Following [12] and [13], we define the ratio of the linear natural frequencies, $\mu \equiv \omega_m/\omega_e$, as the natural singular parameter of the coupled system. In the sequel, after scaling the time, $\tau \equiv \omega_m t$, and using the following state or phase space transformation:

$$x_1 = x, \quad x_2 = x', \quad x_3 = \tau, \quad z_1 = \frac{q}{\mu^2}, \quad z_2 = \frac{q'}{\mu} \tag{6}$$

the second order system (5) assumes the autonomous first order description through the following state space formulation:

$$x_1' = x_2$$
$$x_2' = -x_1 - 2\zeta_m x_2 + \epsilon \alpha \mu^2 z_2^2$$
$$x_3' = 1$$
$$\mu z_1' = z_2$$
$$\mu(1 + \alpha x_1)z_2' = -z_1 - 2\zeta_e z_2 - \alpha \mu x_2 z_2 + \hat{E}(x_3) \tag{7}$$

Symbol $()'$ denotes time derivative with respect to the dimensionless time scale τ and variable $\hat{E}(x_3)$ denotes the properly scaled voltage excitation. The right-hand side of Eq. (7) and (2) represents the mechanical restoring force $g(x_1, x_2, z_2)$. The uncoupled natural frequencies are 1 and $1/\mu$ and we define the normalized frequency difference $\Delta\omega \equiv (1-\mu)/\mu$ as the spectral gap of the coupled system.

The forced coupled system (5) is described as a GSP of the mechanical oscillator ((5) and (1)). This description offers unparalleled leverages for a systematic global geometric analysis. The GSP formulation decomposes geometrically the five-dimensional phase space of system (7) into regions of slow dynamics and fast dynamics. More precisely, the slow–fast time decomposition is global and sharp when the natural singular parameter is sufficiently small: $0 < \mu < 1$. The region of the phase space containing only slow motions forms a smooth geometric object of three-dimensions. It is called the Slow Invariant Manifold (SIM) since a motion originating on it remains on it for all times provided that it is global. Once the SIM is determined we have also determined automatically the complementary fast invariant manifolds.

The Limiting Slow and Fast Dynamics
A global slow–fast time decomposition is essentially substantiated by examining closely the limiting system when $\mu \to 0$. The singularly perturbed system (7) is continuous in the singular parameter μ. At $\mu = 0$ its essential dimension drops smoothly from five to three. The resulting system is simply the linear mechanical oscillator representing the zero order reduced slow system:

$$x_1' = x_2$$
$$x_2' = -x_1 - 2\zeta_m x_2$$
$$z_1 = \hat{E}(\tau), \quad z_2 = 0 \tag{8}$$

After rescaling the time as $\tilde{t} = \tau/\mu$, we obtain the fast time scale description of the FOS (7). At $\mu=0$ its dimension drops from five to two. This is the zero order reduced fast system and it is given by the electrical oscillator:

$$z_1' = z_2$$
$$(1 + \alpha x_1) z_2' = -z_1 - 2\zeta_e z_2 + \hat{E}(\tilde{\tau})$$
$$x_1 = x_1(0), \quad x_2 = x_2(0) \tag{10}$$

For $\mu \neq 0$, it is clear that the slow dynamics (8) can interact with the fast electrical oscillator dynamics (9) due to the fact that the mechanical oscillator displacement x_1 appears on the left-hand side of the reduced fast system (9). Slow–fast interaction will occur for any value of the singular parameter when the voltage excitation is close to the linear natural frequency of the electrical oscillator. We conjecture that this interaction will become strong and irregular (chaotic since we deal with a non-linear system) when the spectral gap $\Delta\omega$ narrows.

The zero order reduced slow and fast systems provide the seeds for the corresponding reduced systems of the FOS. For non-zero value of the singular parameter μ, it is expected to have reduced order slow and fast systems for the FOS (7). Physically this means that we should be able to observe slow and fast motions in the FOS.

THE CASE OF HARMONIC VOLTAGE EXCITATION

Having laid the geometry of the slow–fast time decomposition, we now restrict our study when the coupled system is excited, when at rest, by the harmonic voltage:

$$\hat{E}(\tau) = \beta \sin(\Omega\tau)$$

where β and Ω denote amplitude and frequency (rad/s), respectively. We present some exploratory direct numerical simulations of the FOS (7) with parameters listed in Table 1. When setting the singular parameter $\mu = 0.01$, we get a weakly coupled system with spectral gap $\Delta\omega = 99$. The amplitude of the excitation is 1.00 and the forced response at the three frequencies,

0.5, 95.0, 50.0 rad/s, is considered; the initial conditions are set to zero. First, we consider the voltage excitation with frequency Ω=0.5 rad/s for which we expect resonance since the linear oscillator is forced internally by the squared current quadratic non-linearity. Although the system is non-linear we call this a linear resonance because the slow system is approximated by a linear system resonating at 0.5 rad/s Eq. (24). This is a quite slow time scale excitation; Fig. 2a, b reveals that both the mechanical and electrical oscillators respond in a slow time scale. Now the existence of a SIM implies that this slow time scale response holds for a wide range of forcing parameters. Below this shall be justified by analytic computation of the SIM. Next, we consider the excitation at frequency 95.0 rad/s which is close to the linear frequency of the electrical oscillator (100 rad/s). We have excitation in the intrinsic fast time scale of the system. Fig. 2c, d reveals that the mechanical oscillator responds with very slow dynamics despite the fact it is excited with a high frequency excitation. On the other hand, the electrical oscillator responds with fast oscillations modulated with a slowly varying envelope. The envelope seems to have a frequency close to that of the mechanical oscillator. Here we see in a clear manner how the slow dynamics interact with the fast; the dynamics of the electrical oscillator do not reside on a SIM whereas the dynamics of the mechanical oscillator might reside on it. Finally we present a typical case with intermediate frequency at 50.0 rad/s. Fig. 2e, f reveals a sharp separation of slow and fast oscillations. The slow ones are confined to the mechanical oscillator and the fast ones to the electrical one. The conclusion is the fact that in all cases the mechanical oscillator responds with slow dynamics. Below we proceed to compute the SIM and examine whether the slow reduced system predicts the slow dynamics.

Table 1. Electromagnetic system with $\Delta\omega$=99.

Parameter	Symbol	Value
Singular parameter	μ	0.01
Mechanical dissipation factor	ζm	0.01
Electrical dissipation factor	ζe	0.1
Electro-mechanical coupling	ϵ	5
Inductance non-linearity	α	10

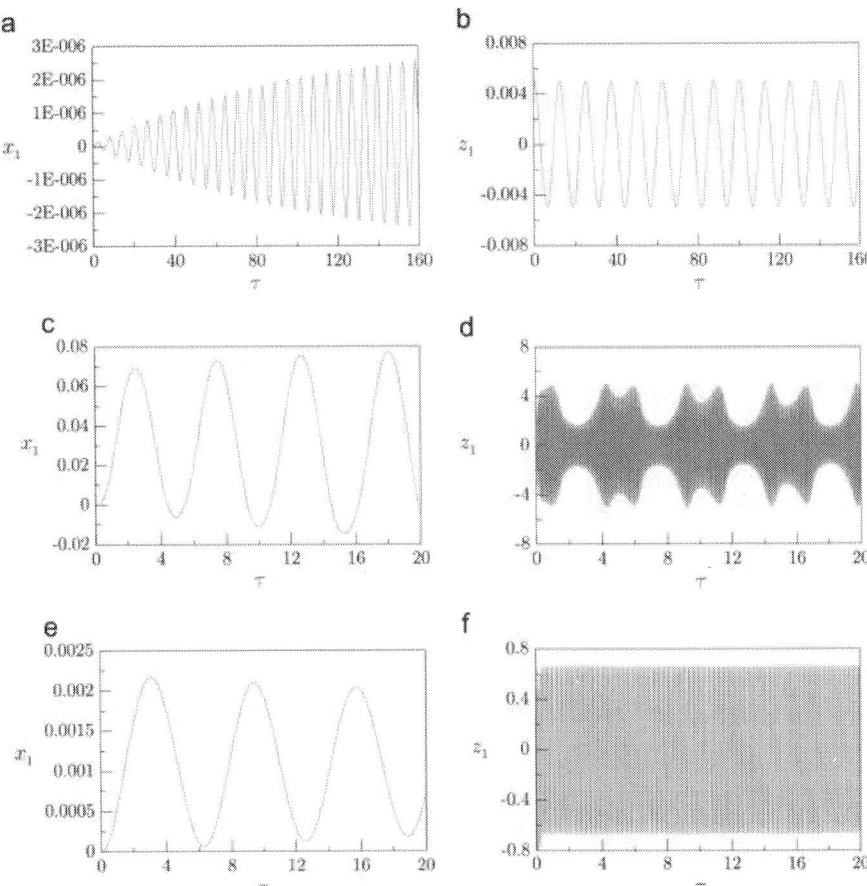

Figure 2. Voltage excitation response of the coupled system: (a), (b) Ω=0.5 rad/s; (c), (d) Ω=95 rad/s; (e), (f) Ω=50 rad/s.

ANALYTIC COMPUTATION OF THE SIM

We shall compute analytically the SIM for the generic case of slow voltage excitation while the mechanical external force is set to zero. The results will be restricted to the subcase of harmonic excitation. To carry out the analytic computations effectively in an algorithmic setting, the FOS (7) is rewritten compactly as follows:

$$X'=AX+\mu^2 P_1(Z)Z+Q\mu C(X)Z'=BZ-\mu P_2(X)Z+G(X) \tag{10}$$

where the state of the system is separated into the slow and fast sub-states:

$$\mathbf{X}=[x_1,\ x_2,\ x_3]^T,\quad \mathbf{Z}=[z_1,\ z_2]^T \tag{11}$$

The matrices entering Eq. (10) are defined as follows:

$$\mathbf{A}=\begin{pmatrix} 0 & 1 & 0 \\ -1 & -2\zeta_m & 0 \\ 0 & 0 & 0 \end{pmatrix},\quad \mathbf{P}_1(\mathbf{Z})=\begin{pmatrix} 0 & 0 \\ 0 & \alpha z_2 \\ 0 & 0 \end{pmatrix},\quad \mathbf{Q}=\begin{pmatrix} 0 \\ 0 \\ 1 \end{pmatrix}$$

$$\mathbf{B}=\begin{pmatrix} 0 & 1 \\ -1 & -2\zeta_e \end{pmatrix},\quad \mathbf{C}(\mathbf{X})=\begin{pmatrix} 1 & 0 \\ 0 & 1+\alpha x_1 \end{pmatrix}$$

$$\mathbf{P}_2(\mathbf{X})=\begin{pmatrix} 0 & 0 \\ 0 & \alpha x_2 \end{pmatrix},\quad \mathbf{G}(\mathbf{X})=\begin{pmatrix} 0 \\ \beta\sin(\Omega x_3) \end{pmatrix} \tag{12}$$

For sufficient small coupling μ the existence of the SIM is guaranteed by the Global Center Manifold Theorem [14]. Our system differs from the systems for which the manifold theorem is stated since the slow variable $\mathbf{X}$ appears on the left side of Eq. (10-2). Given the smoothness of the dynamical systems, it is reasonable to assume that at least part of the SIM admits an analytic description via the graph of a smooth non-linear function of the slow sub-state:

$$\mathbf{Z}=\mathbf{H}\mu(\mathbf{X})=[\mathbf{H}\mu1(x1,x2,x3),\mathbf{H}\mu2(x1,x2,x3)]^T \tag{13}$$

We identify the manifold with the graph w_μ of function (13) defined to be the set of points:

$$\mathbf{W}\mu\equiv\{(\mathbf{X,Z})\in R^5:\mathbf{Z}-\mathbf{H}\mu(\mathbf{X})=0\} \tag{14}$$

Manifold description (13) accounts for the quite natural expectation that the reduced slow dynamics may stem as a regular perturbation of the uncoupled slow system, i.e. equation (8), due to its coupling-induced extension by the fast oscillator. This particular manifold description has been used in several investigations to isolate the slow scale dynamics of coupled systems in mechanics and biology. In solid mechanics, manifold description (13) has been used to compute analytically global non-linear normal modes in coupled systems of oscillators. For example, reference [15] discusses how fundamental is to view the generic coupled structural–mechanical oscillator system as a singular perturbation problem and compute the underlining slow and fast invariant manifolds. Moreover, invariant manifold description (13)has been used to compute local non-linear normal modes of vibration in vibratory systems [16]. In biology, manifold description (13) is used to compute slow manifolds in neuron oscillator models and in particular to correlate the appearance of fast bursting dynamics to the folding of a slow manifold; in this case it needs to be treated properly in order to capture manifold folding [17]. In this respect it is worth remarking that the mechanical and biological systems studied as 'slow–fast' and admitting the classical singular perturbation description are qualitatively approximated by the zero order slow system; here we have an interesting case where this does not hold.

The graph description yields a manifold that is non-linear in the sense that it does not admit a span representation by linear subspaces. Embedded in the five-dimensional Euclidean space, it has locally the structure of a linear three-dimensional Euclidean space. The actual SIM of our system might have additional features such as a folding in parameter space. In that case the graph representation via a single-valued function is not adequate.

Now the manifold w_μ is an invariant of the motion if and only if the vector field of the dynamical system (10)at every point on the manifold resides in the tangent space passing through that point. The collection of all tangent spaces is called the tangent bundle space in the language of differential geometry [18] and [19]. Restating, the manifold as a geometric entity is invariant if and only if the vector field of the dynamical system resides in the tangent bundle space of the graph (14). In mathematical language this global tangency requirement is expressed as a partial differential equation:

$$\mu C(X) D_X H_\mu [AX + \epsilon \mu^2 P_1(H_\mu) H_\mu + Q] = BH_\mu - \mu P_2(X) H_\mu + G(X) \quad (15)$$

where term $_{DX}$ denotes the first order differential operator. This is called the Invariant Manifold Condition (IMC). Its solution gives the manifold function $_{H\mu}$. We expect that for sufficiently small coupling μ the SIM manifold is exponentially stable over a global region of the phase space since our system is dissipative. The stability stems from the fact that the zeroth order fast system (9) is a linear dissipative system with a single equilibrium state and thus all its initial conditions are attracted by the forced (finite) steady state solutions. An approximate solution to the IMC is computed by expanding the manifold function as a finite series in the coupling:

$$H_\mu(X) \simeq H_\mu^{[N]}(X) = \sum_{n=0}^{N} \mu^n H_n(X), \quad H_n = \begin{bmatrix} H_{1n} \\ H_{2n} \end{bmatrix} \quad (16)$$

For convenience, term $_{Hn}$ is referred to as the n -th order approximation term and term $H_\mu^{[N]}$ is referred to as the N-th approximate manifold. The components of the approximation terms must satisfy the following compatibility relation:

$$H_{2n} = H'_{1(n-1)}, \quad n = 1, \dots, N \quad (17)$$

This relation guarantees that the scaled current component is the time derivative of the scaled charge component when the dynamics reside on the slow manifold. It is a useful relation as a diagnostic tool in detecting erroneous terms in the analytic computations of the manifold function solution reported in the Appendix.

SLOW DYNAMICS OF THE FULL ORDER SYSTEM

We will study numerically, with the help of an existing SIM, the steady state forced dynamics for the case where the external mechanical forcing at the mass is zero and the external harmonic voltage introduced in Section 4, namely $\hat{E}(\tau) = \beta \sin(\Omega \tau)$, is applied. Numerical experiments in combination with the computed approximate manifolds will verify the existence of a stable SIM, which, in turn, will be used to interpret the forced response of the FOS for a wide range of forcing parameters. A systematic way to explore the qualitative dynamics of a non-linear system is to compute attractor diagrams of the forced response over the space of parameters of interest. Here this is done by applying the dimension reduction method of Poincaré sections or mappings [18] and [19]. We take Poincaré global sections of the system by simply sampling the numerically computed state (4th order Runge–Kutta with adaptive time step) at a quite large number of equally spaced time points, the time interval being set to the forcing period T:

$$
\begin{aligned}
&x_{1k} \equiv x_1(\tau + kT), \quad x_{2k} \equiv x_2(\tau + kT), \\
&z_{1k} \equiv z_1(\tau + kT), \quad z_{2k} \equiv z_2(\tau + kT), \\
&k = 1, 2, \ldots, K_L, K_L + 1, \ldots, K_R - 1, K_R
\end{aligned}
\tag{18}
$$

where $\{_{KL}+1, \ldots, _{KL}+n, _{KR}\}$ gives the tail of the iteration. As it is well known that the above sequences signify the iterations of the first return map [18] and [19]. Thus the limit of a long iteration gives a fixed point of the Poincaré mapping. For example, a fixed point represents a period-1 attractor and a wandering iteration signifies either a chaotic or a quasi-periodic attractor.

Now we compute the following limiting iteration-sets:

$$
\begin{aligned}
&A_{x_1}(\beta, \Omega) \equiv \{x_{1k}\}_{k=K_L}^{K_R}, \quad A_{x_2}(\beta, \Omega) \equiv \{x_{2k}\}_{k=K_L}^{K_R}, \\
&A_{z_1}(\beta, \Omega) \equiv \{z_{1k}\}_{k=K_L}^{K_R}, \quad A_{z_2}(\beta, \Omega) \equiv \{z_{2k}\}_{k=K_L}^{K_R}
\end{aligned}
\tag{19}
$$

The tail of the iteration furnishes attractor diagrams over the space of the forcing parameters. For fixed voltage amplitude at $\beta^\square$ we compute response amplitude–forcing frequency diagrams by sweeping quasi-statically the forcing frequency Ω . For fixed voltage frequency $\Omega^\square$, we compute response amplitude–forcing amplitude diagrams by sweeping quasi-statically the forcing amplitude.

In the numerical investigations that follow the system parameters are set as reported in Table 2. Fig. 3a, b, c shows typical response amplitude-sweep frequency attractor diagrams $_{Ax1}(\beta^\square,\Omega)$ for quasi-static forward and backward frequency sweeps for $\mu=0.01$, $\mu=0.1$ and $\mu=0.2$, respectively. For $\mu=0.01$ the system behaves almost linearly, so that the forward sweep coincides with the backward one, and, as shown by the zoomed view in Fig. 3a, it reveals a stable period-1 attractor which resonates around $\Omega=0.5$ in the frequency interval $(0, 5]$. Note that low linear resonant frequency is at 1. In Fig. 3b, the different branches corresponding to forward and backward sweeps for $\mu=0.1$ show how the coupling non-linearity affects the dynamics; in particular, the upper branch of the period-1 attractor is obtained from the backward sweep. This branch arises from the lower bound of the chaotic region located in the frequency interval $4.0 \leq \Omega \leq 4.5$ and then, as the frequency decreases, it grows until the instability is reached around $\Omega=1.55$; furthermore, the resonance around $\Omega=0.5$ persists. For $\mu=0.2$ the forward sweep in Fig. 3c reveals a stable period-1 attractor which again resonates around $\Omega=0.5$. The period-1 attractor varies smoothly without qualitative changes up to a critical frequency where it suffers a finite jump and thus it is placed on another higher amplitude branch which continues smoothly without qualitative changes up to the end of the interval of interest. The closer views referred to the case with $\mu=0.2$, shown in Fig. 4, pinpoint a critical frequency near $_{\Omega A}=1.633$. The backward frequency sweep produces a completely different path as shown in Fig. 4. At the critical frequency $_{\Omega B}=1.546$ the period-1 attractor changes rather abruptly to a chaotic one, then, as the frequency decreases, the chaotic attractor changes smoothly into a period-1 attractor. At the critical frequency $_{\Omega C}=0.966$ the period-1 attractor suffers an abrupt downwards finite jump. Before the jump the amplitude of the period-1 attractor is about two orders of magnitude higher than the amplitude at the linear resonance $\Omega=0.5$. These steady state dynamics evolve at a slow time scale set by the low frequency of the voltage excitation and two questions arise. As far as the dimension and the time scale are concerned, are these dynamics unified by a Slow Invariant Manifold? Which oscillator, the mechanical or the electrical, is the source of the slow time scale dynamics?

Table 2. Electromagnetic system with $\Delta\omega=4$.

Parameter	Symbol	Value
Singular parameter	μ	0.01, 0.1, 0.2
Mechanical dissipation factor	ζm	0.01
Electrical dissipation factor	ζe	0.1
Electro-mechanical coupling	ϵ	5
Inductance non-linearity	α	10

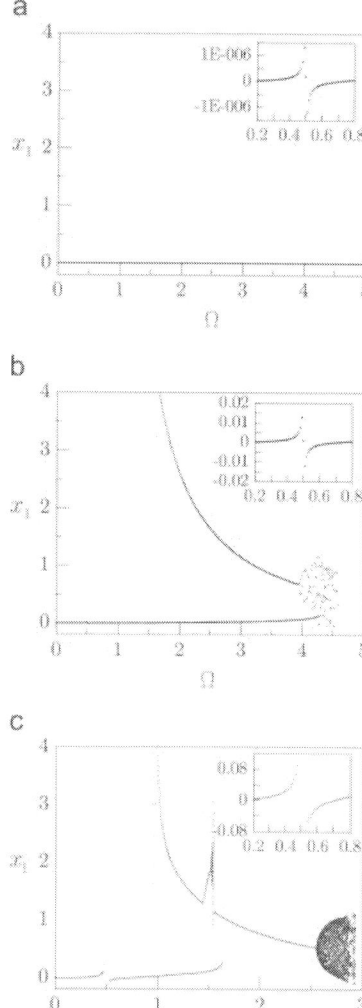

Figure 3. Forced response amplitude–frequency sweep plot of the FOS for $\beta^-=1$: (a) $\mu=0.01$, (b) $\mu=0.1$, (c) $\mu=0.2$.

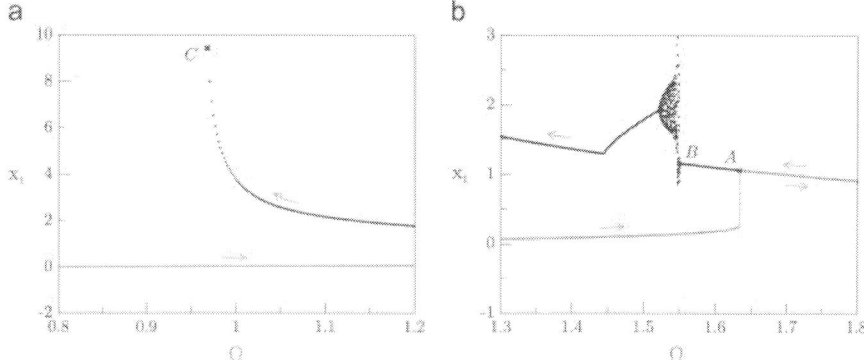

Figure 4. Details of the forced response amplitude–frequency diagram of the FOS in the neighborhood of jump frequencies for μ=0.2: forward (grey) and backward (black) sweeps.

Fig. 5 shows an amplitude–frequency diagram over the frequency range (0, 3]. We see very clearly the behavior of the FOS at various levels of voltage amplitudes. The behavior is robust and varies smoothly over the examined range of forcing amplitudes. We see that the FOS goes into a chaotic regime for frequencies higher than 2.0. The forward sweep reveals the persistent non-linear resonance phenomenon as the frequency approaches quasi-statically the value 1. Note that near 1.5 we have an abrupt change from periodic into chaotic behavior. The jump observed at Ω_B=1.546 entails oscillations of the mechanical mass about an average amplitude value which increases by about an order of magnitude. As shown in Fig. 6, such oscillations around a non-zero average value are due to the current squared non-linearity and the jump reveals a pull-in phenomenon. It is worth mentioning that the frequency of the mechanical oscillator is twice that of the forced electrical part. This relation does not change in view of the jump.

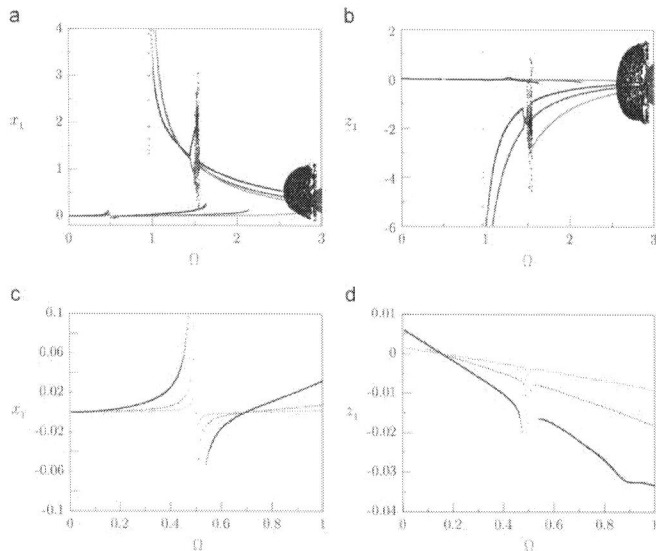

Figure 5. Frequency–amplitude plots for the FOS for different voltage amplitude β: 1.0 (black), 0.5 (dark grey) and 0.25 (light gray): (a) mechanical displacement (x_1); (b) electrical displacement (z_1); (c), (d) zoomed views of x_1 and z_1 for $0 \leq \Omega \leq 1$.

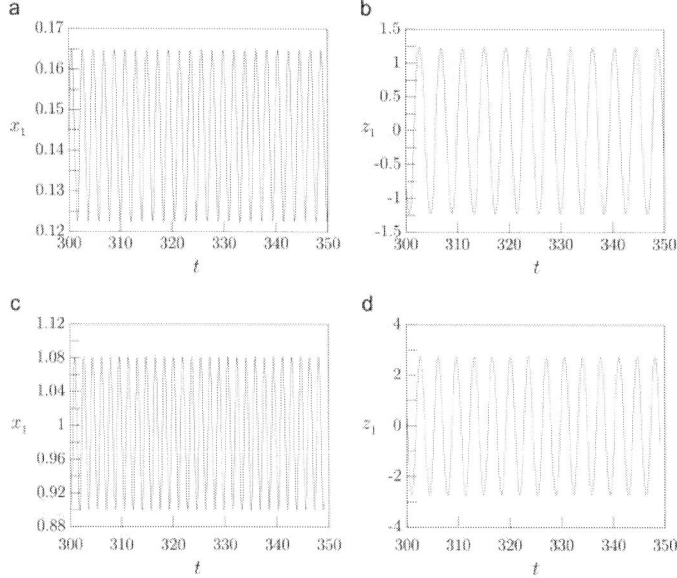

Figure 6. Forced mechanical and electrical response of the FOS for a forcing frequency below and above the jump at Ω_B: (a), (b) $\Omega = 1.5$ and (c), (d) $\Omega = 1.8$.

THE REDUCED ORDER SLOW SYSTEM

We will use the reduced order slow system to understand the slow dynamics of the FOS in a typical frequency-amplitude diagram as the one shown in Fig. 3. The importance of the existence of a SIM resides into the fact that the slow time scale behavior is viewed spherically by geometric means and thus it is much more informative than individual quantitative analysis. We assume the existence of a stable slow invariant manifold. This will be verified in the course of comparing the dynamics of the FOS to the dynamics predicted by the approximate slow manifolds. Now the restriction of the FOS onto the SIM gives the reduced order slow system:

$$
\begin{aligned}
x_1' &= x_2, \\
x_2' &= -x_1 - 2\zeta_m x_2 + \epsilon\alpha\mu^2 H_{2\mu}^2(x_1, x_2, x_3), \\
x_3' &= 1,
\end{aligned}
\tag{20}
$$

$$
z_1 = H_{1\mu}(x_1, x_2, x_3), \quad z_2 = H_{2\mu}(x_1, x_2, x_3)
\tag{21}
$$

The first equation gives the master slow dynamics. The second equation gives the slave slow dynamics, which are the stretched electric charge and current. Below we analyze approximates of the reduced slow system (21) by using the computed approximations of the slow manifold. By the term N-th order slow system we mean the approximate reduced slow system stemming from the restriction of the FOS onto the computed N-th approximate slow manifold. As mentioned above, the zero order slow system(8) is the limit of the FOS as $\mu \to 0$. Also from a pure geometrical standpoint, it is the restriction of the FOS on the zeroth approximate slow manifold. For convenience, we rewrite it

$$
x_1' = x_2, x_2' = -x_1 - 2\zeta m x_2
\tag{22}
$$

$$
z_1 = H_{10}(\tau) = \hat{E}(\tau) = \beta \sin(\Omega\tau), \quad z_2 = H_{20}(\tau) = 0
\tag{23}
$$

We see that the approximate master slow system coincides with the unforced uncoupled mechanical oscillator. Clearly the zeroth order slow system predicts only the slow dynamics stemming from the free dynamics disturbances of the mechanical oscillator. Given the existence of a SIM, the predictions of the zeroth order slow system hold for a FOS with a sufficiently large spectral gap and excited simultaneously by free slow dynamics and slow voltage excitation. The steady state response due to voltage excitation is very small when the coupling is weak so that, at the zeroth order level, the approximate electric charge and current are identically zero, this being a singular case. Fig. 3 reveals that when the quasi-statically sweeping frequency approaches the frequency $\Omega = 0.5$ the mechanical displacement amplitude increases and it undergoes a phase shift after crossing this critical frequency. This is the linear resonance phenomenon occurring at 1/2 the natural frequency of the mechanical oscillator and this is due to the quadratic non-linearity so that the zero order slow system does not predict this response. Thus the next order of the approximate slow system is considered. It turns out that, due to the peculiarity of our system, the next order approximate slow system which contains explicitly μ^2 term is the fourth one:

$$x_1' = x_2$$
$$x_2' = -x_1 - 2\zeta_m x_2 + \epsilon\alpha\mu^4\beta^2\Omega^2 \cos^2(\Omega\tau) \tag{24}$$

$$z1 \simeq H10(\tau) = \beta\sin(\Omega\tau) z2 \simeq H10(\tau) + \mu H21(\tau) = \mu\beta\Omega\cos(\Omega\tau) \tag{25}$$

Note that the fourth order slow system is purely linear but it captures accurately the resonance of the FOS at the forcing frequency $\Omega = 0.50$ (see Fig. 3). Thus the first order approximate slow invariant manifold reveals very clearly the mechanism by which the FOS resonates at 1/2 of the natural frequency of the linear oscillators. This mechanism is the quadratic electric current non-linearity which appears as a squared forcing term in the fourth order slow system.

The fourth order system can only approximate rather well the lower branch provided that the amplitude is sufficiently small so that the linear terms dominate (see Fig. 3). To better understand the bifurcations in the frequency-amplitude diagram shown in Fig. 3, we computed higher order slow systems. The jumps in Fig. 3clearly indicate that something happens in the geometry of the SIM. The zeroth order and the fourth order slow systems verify the existence of the SIM in a restricted region of parameters. We would also like to see if the higher order slow systems converge in the region where the fourth order system captures the FOS behavior. Moreover, the higher order terms are expected to provide better approximations in capturing the response non-linear features.

On the basis of the computed first order approximate SIM, the slow charge and current approximations are given by

$$q \simeq \mu^2 H_{10}, \quad q' \simeq \mu H_{21} \rightarrow q = \mu^2\beta\sin(\Omega\tau), \quad q' = \mu^2\beta\Omega\cos(\Omega\tau) \tag{26}$$

We have the interesting result that when the electric circuit is forced at low frequencies including the natural frequency of the mechanical oscillator the circuit responds in a slow time scale and the current is proportional to the first derivative of the voltage source. This result is obtained elegantly by computing analytically the SIM.

The Higher Order Slow Systems

Since the 4th order system is linear, we have computed higher order approximate slow systems in an effort to explain the features of the response amplitude-sweep frequency diagram of the FOS shown in Fig. 3 and Fig. 4. To discuss the properties of the higher order approximate slow systems, we present the computed 10th order approximate reduced slow system:

$$x_1{}' = x_2$$

$$
\begin{aligned}
x_2{}' = & -x_1 - 2\zeta_m x_2 + \epsilon\alpha\mu^4 H_{21}^2 + 2\epsilon\alpha\mu^5 H_{21}H_{22} \\
& + \epsilon\alpha\mu^6\left[H_{22}^2 + 2H_{21}H_{23}\right] + 2\epsilon\alpha\mu^7[H_{21}H_{24} + H_{22}H_{23}] \\
& + \epsilon\alpha\mu^8\left[H_{23}^2 + 2H_{22}H_{24} + 2H_{21}H_{25}\right] \\
& + 2\epsilon\alpha\mu^9[H_{23}H_{24} + H_{22}H_{25} + H_{21}H_{26}] \\
& + \epsilon\alpha\mu^{10}\left[H_{24}^2 + 2H_{27}H_{21} + 2H_{26}H_{22} + 2H_{23}H_{25}\right]
\end{aligned}
$$

$$x_3{}' = 1 \tag{27}$$

The above is computed by using the 7th order approximation terms to IMC (15):

$$z_1 \simeq \hat{H}_{1\mu}(x_1, x_2, \tau) = \sum_{k=0}^{6} \mu^k H_{1k}(x_1, x_2, \tau)$$

$$z_2 \simeq \hat{H}_{2\mu}(x_1, x_2, \tau) = \sum_{k=0}^{7} \mu^k H_{2k}(x_1, x_2, \tau) \tag{28}$$

On the basis of these approximations, the slow electric charge and current are predicted to be approximately:

$$q \simeq \mu^2 \sum_{k=0}^{6} \mu^k H_{1k}(x_1, x_2, \tau), \quad q' \simeq \mu \sum_{k=0}^{7} \mu^k H_{2k}(x_1, x_2, \tau) \tag{29}$$

At this point we discuss the mathematical properties of the higher order approximate reduced slow systems. This discussion will shed light into the slow behavior of the FOS captured by the hierarchy of the computed approximate slow systems. It turns out that the 5th order reduced system is purely linear. However it has a quadratic non-linear dependence on the voltage waveform which is represented by the x_3 variable.

The 6th and 7th order approximate reduced slow systems contain pure linear terms in x_1 and x_2 which are very small perturbations to the linear terms of the mechanical oscillator. Thus, we do not expect any qualitative change. The 8th and 9th order approximate reduced systems add the quadratic non-linear terms: $x_1^2, x_2^2, x_1 x_2$. Finally the 10th order approximate reduced system adds the cubic non-linear terms: $x_1^3, x_2^3, x_1 x_2^2$. Depending on the sign of these quadratic and cubic non-linearities the reduced order system can give rise to the sought qualitative changes of the dynamics. It is worth emphasizing that the latter non-linearities stem from the restriction of the FOS to the SIM and are not directly related to the actual FOS quadratic non-linearities.

We compare the frequency-amplitude diagrams of the 4th, 8th, and 10th order approximate reduced slow systems to that of the FOS, shown in Fig. 7. On one hand, the computed 4th, 8th, 10th order approximate slow systems reproduce faithfully the forward branch of period-1 attractors up to the critical frequency$_{\Omega A}$=1.633 where the jump occurs. On the other hand, unexpectedly, the computed 10th order approximate slow system captures faithfully the FOS up to the point of jump $_{\Omega C}$=0.966 from the resonance condition on the backward period-1 attractor branch. Recall that this jump was detected by sweeping backward the frequency. In other words, the range where the slow systems below the 9th order and 10th agree shrinks to frequencies below frequency 1. From a mathematics point of view, we have the rather interesting result that hierarchical approximate systems undergo qualitative changes as the order is increased. The qualitative change is the infinite instability. It is worth noticing that this qualitative change occurs close to the jump parameters. We believe that these qualitative changes are not mathematical artifacts. Moreover there is a region over which all orders of approximate reduced slow systems converge and predict exceptionally well the slow dynamics of the FOS. The existence of the SIM is verified by the fact that all approximate slow manifolds predict the behavior of the full order system in the frequency range (0, 1.633). The SIM depends parametrically on the forcing parameters. In Fig. 8 the mechanical oscillator restoring force g$_{(x1,x2,z2)}$, as defined in Section 3, is plotted against the applied voltage $\hat{E}(\tau)$ for β=2.3; while the lower order models, 4th and 8th, are able to reproduce the FOS restoring force only on average, the 10th order one captures more accurately the FOS restoring force behavior in the voltage amplitude range between −1 and 1.

CONJECTURE

The three-dimensional stable SIM undergoes a folding at the detected points of jump frequencies. The folded part of the manifold gives rise to a multi-valued function and generates the mechanisms of jump and hysteresis. At the jumps the high order approximate reduced slow systems inherit the folding feature and exhibit an unbounded instability giving rise to an infinite jump. We will assume that we have a SIM with complicated structure between the jump frequencies but with a clear structure up to the jump in the forward sweep. It is logical to propose that this complicated manifold structure is responsible for the behavior of the FOS in the frequency domain (0.9, 1.7). *The approximate slow systems predict exceptionally well the forward behavior of the FOS in the region* (0, 1.5) *where the SIM admits a single-valued function representation.*

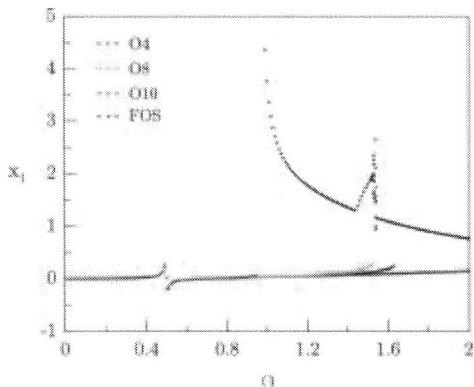

Figure 7. Forced response amplitude–frequency sweep plot of the FOS. Comparison with the approximate slow systems.

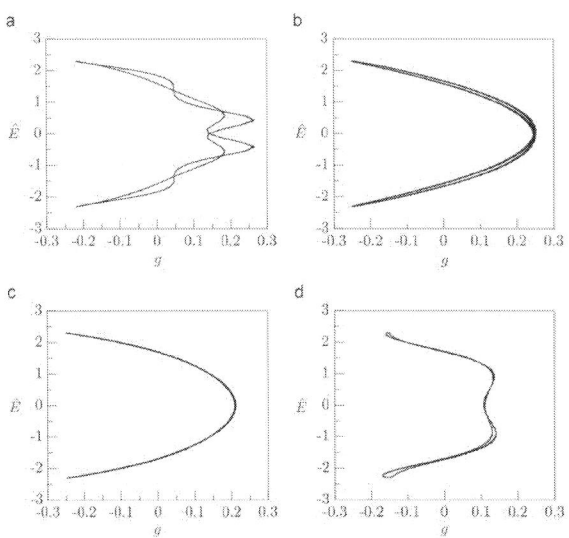

Figure 8. Applied voltage vs. restoring force for β=2.3 and Ω=0.6: (a) FOS; (b) 4th order; (c) 8th order; (d) 10th order.

COMPLEX CHAOTIC INTERACTION DYNAMICS

We present a numerical experiment of the forced dynamics of the coupled system (7) with spectral gap$\Delta\omega$=4 and parameters as listed in Table 2. The forcing frequency is fixed at Ω=0.6 and we sweep the voltage amplitude quasi-statically over the range [0, 2]. Fig. 9 shows the behavior of the FOS at two different coupling levels, namely μ=0.2, μ=0.15. As the coupling level is lowered the region before the jump extends to the right. As expected, Fig. 9b shows that the higher order approximate slow systems provide a better prediction of the FOS than the low order approximate linear slow systems within the region of forcing parameters corresponding to period-1 attractors. Next, the scenario of the full system is expanded up toβ=10 in the bifurcation diagrams presented in Fig. 10. They refer to both the mechanical displacement and the electric charge and show that the initial periodic attractor changes into higher order and quasi-periodic attractors as the forcing amplitude increases. Moreover, initial chaotic attractors characterized by a limited amplitude response appear below β=6. Then, a new periodic attractor arises around β=6.9 after which a dramatic qualitative change in the response occurs. By comparing the mechanical and the electric bifurcation diagrams, Fig. 10a and b, respectively, a qualitative agreement can be observed, showing that the electric part (slave) follows the mechanical one (master). The alternating pattern of regular and irregular dynamics due to non-linearity is governed by the same β critical values. However above β=6.9 the mechanical oscillator shows a jump leading to irregular oscillations around a higher mean value whereas the jump is absent in the electric response.

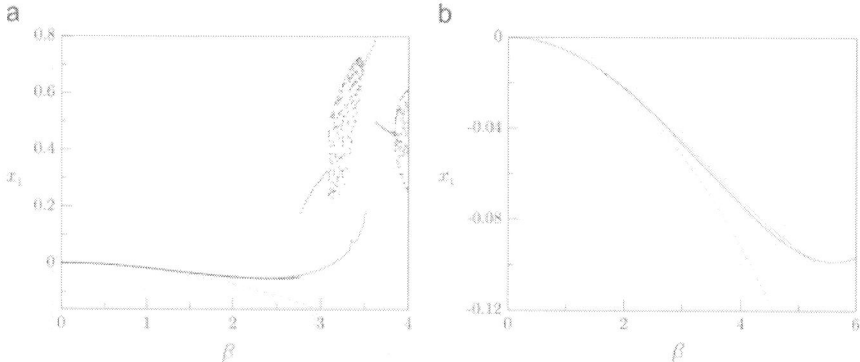

Figure 9. Bifurcation diagrams at two different coupling levels. Comparison between full system (solid), 4th (dashed) and 8th (dotted) order SIM numerical bifurcation diagrams for Ω=0.6: (a) μ=0.2 and (b) μ=0.15.

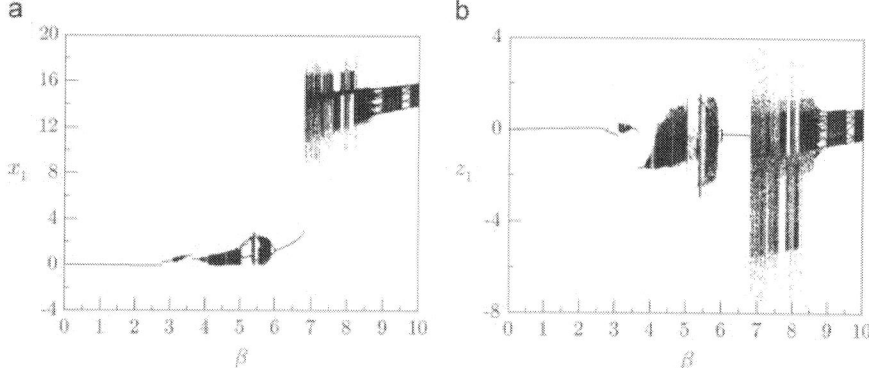

Figure 10. Bifurcation diagrams: (a) mechanical displacement and (b) electric charge.

CONCLUSIONS

The dynamics of a coupled non-linear electro-magneto-mechanical system was addressed by means of analytical and numerical approaches. When the oscillator is forced via harmonic voltage excitation of the electric circuit the long term mechanical dynamics evolves in a purely slow timescale. Therefore a slow invariant manifold was analytically derived and numerically validated by means of experiments on both full- and reduced-order systems. The slow time scale forced frequency–amplitude response of the full order system was computed by means of Poincaré mappings. The latter were analyzed in depth by using approximate slow invariant manifolds and various order reduced slow systems. Three interesting non-linear phenomena were observed. The non-linear resonance related to the current quadratic non-linearity which imposes a natural linear resonance at half the frequency of the linear oscillator. The pull-in phenomenon denoted by a jump after which the mass of the linear mechanical oscillator is pulled at a large distance and it is forced to oscillate about it. A peculiar irregular dynamics for high excitation amplitude involving a number of bifurcations characterized by dramatic qualitative changes of both the mechanical and electrical responses.

Further investigations will be directed to explore more in detail crucial aspects such as the mechanism of energy transfer between the subsystems and the overall behavior in the presence of different sources of excitation.

REFERENCES

1. R. Bishop, The Mechatronics Handbook, CRC Press, Boca Raton, 2006.
2. H. Barth, Sensors and Sensing in Biology and Engineering, second ed., Springer-Verlag, Austria, 2003.
3. M.W. Ashaf, S. Tayyaba, N. Afzulpulkar, Micro electromechanical systems based microfluidic devices for biomedical applications, Int. J. Mol. Sci. 12 (2011) 3648–3704.
4. R.M. Lin, W.J. Wang, Structural dynamics of microsystems – current state of research and future directions, Mech. Syst. Signal Process. 20 (2006) 1015–1043.
5. S.D.A. Hannot, D.J. Rixen, Building and reducing a three-field finite-element model of a damped electromechanical actuator, J. Microelectromech. Syst. 20 (2011) 665–675.
6. R. Yamapi, P. Woafo, Dynamics and synchronization of coupled self-sustained electromechanical devices, J. Sound Vib. 285 (2005) 1151–1170.
7. J. Rhoads, S.W. Shaw, K.L. Turner, Nonlinear dynamics and its applications in microand nanoresonators, J. Dyn. Syst. Meas. Control 132 (2010) 034001–034014.
8. M.I. Younis, MEMS Linear and Nonlinear Statics and Dynamics, Springer, New York, 2011.
9. P.S. Landa, Nonlinear Oscillations and Waves in Dynamical Systems, Kluwer Academic Publishers, Dordrecht, The Nederlands, 1996.
10. R. Darula, S. Sorokin, On non-linear dynamics of a coupled electro-mechanical system, Nonlinear Dyn. 70 (2012) 979–998.
11. C. Jones, Geometric singular perturbations in dynamical systems, in: Springer Lecture Notes in Mathematics, vol. 1609, 1995, pp. 22–120.
12. I.T. Georgiou, A.K. Bajaj, M. Corless, Invariant manifolds and chaotic vibrations in singularly perturbed nonlinear oscillators, Int. J. Eng. Sci. 4 (1998) 431–458.
13. I.T. Georgiou, On the global geometric structure of the dynamics of the elastic pendulum, Nonlinear Dyn. 18 (1999) 51–68.
14. J. Carr, Applications of Center Manifold Theory, Springer-Verlag, New York, 1981.
15. I.T. Georgiou, I.B. Schwartz, The slow invariant manifold of a conservative pendulum–oscillator system, Int. J. Bifurc. Chaos 06 (1996) 673–692.
16. S.W. Shaw, C. Pierre, Normal modes for nonlinear vibratory systems, J. Sound Vib. 164 (1993) 85–124.
17. J.M. Ginoux, B. Rossetto, Slow manifold of a neuronal bursting model, in: Understanding Complex Systems, Springer-Verlag, Heidelberg, Germany, 2006.

18. S. Wiggins, Introduction to Applied Nonlinear Dynamical Systems and Chaos, Springer-Verlag, New York, NY, 1990.

19. F. Scheck, Mechanics, Springer-Verlag, Berlin, Germany, 1990.

CITATION

Ioannis T. Georgiou, Francesco Romeo, Multi-physics dynamics of a mechanical oscillator coupled to an electro-magnetic circuit, doi:10.1016/j.ijnonlinmec. 2014.08.007.

CHAPTER 8

Non-Linear Dynamics of a Mechanical System with a Frictional Unilateral Constraint

Giovanni Lancioni[1], Stefano Lenci[1], UgoGalvanetto[2]

[1]Dipartimento di Architettura, Costruzioni e Strutture, UniversitáPolitecnicadelle Marche, Via BrecceBianche 1, 60131 Ancona, Italy
[2]Dipartimento di Costruzioni e Trasporti, Universitá di Padova, Via Marzolo 9, 35131 Padova, Italy

ABSTRACT

We analyze the dynamics of a two-dimensional system constituted by two masses subjected to elastic, gravitational and viscous forces and constrained by a moving frictional mono-lateral surface. The model exhibits a time-varying dynamics capable of reproducing the hopping phenomenon, an unwanted phenomenon observed in many applications such as the motion of a robotic arm on a surface or that of a wiper on a windscreen. The system dynamics, besides being affected by geometrical non-linearities, has a non-smooth nature due to the impact and friction laws involved in the model. The complexity of the resulting equations and of the transition conditions require the problem to be solved numerically. Various periodic motions are found and the effect of varying the system parameters, in particular the friction coefficient, is investigated. Finally, simulations are used to gain some insight the behavior of the windscreen wiper.

INTRODUCTION

The motion of the planar mass–spring–damper system depicted in Figure. 1 is investigated. It consists of two masses m_1 and m_2 which are immersed in a vertical downward gravitational field and are constrained by elastic springs, viscous dashpots and a moving mono-lateral frictional surface. This model, due to [12]where it has been proposed, studied in detail and called frictional impact oscillator, reproduces the hopping phenomenon observed in many applications, such as the motion of a piece of chalk pushed over a blackboard or that of a robotic arm on a plane. The system can also be seen as a simplified model of the transverse section of a wiper blade moving on the windscreen of a car, a problem which originally motivates this work [3], [8] and [23]. The investigation of the dynamics is useful to understand the causes of the hopping and, as a result, to suggest the appropriate remedies to avoid this unwanted motion.

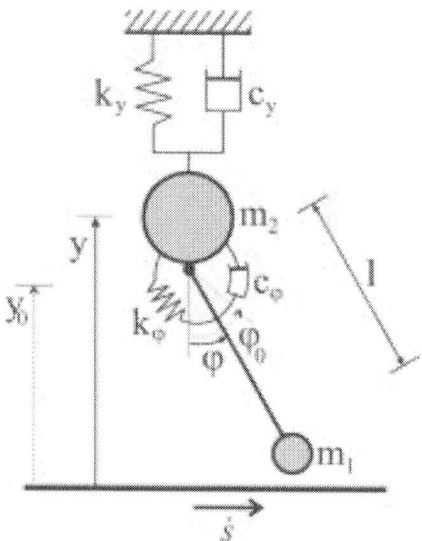

Figure 1. Frictional impact oscillator, see [12].

Despite its apparent simplicity, the model is very intricate due to various non-linearities involved in its dynamics. Large amplitude oscillations of mass m_1, its impacts with the moving surface and friction, acting between m_1 and the moving surface when they are in contact, are all sources of non-linear behavior. Systems with impact and/or friction, as well as stick–slip systems have been the subject of extensive investigations due to the variety of non-linear dynamics phenomena they experience [1], [4], [6], [7], [9],[13], [14], [17] and [20].

The model can be classified as a time-varying system, because its dynamics is divided in three different regimes, each characterized by a different number of degrees of freedom (dof). The free-flight motion occurs when the mass m_1 is detached from the constraining surface, and it is governed by two kinematic parameters (y and ϕ in Figure. 1). When m_1 is in contact we can have slipping or sticking, depending on the value of the friction force. The slip motion is governed by one dof, while in the stick motion there are no dof, and the dynamics is determined by the velocity of the moving surface.

The equations of motion are equipped with appropriate laws of transitions from one regime to the others, which are the key-points of the model, and which require an accurate modeling. In particular, for the impact of m_1 on the surface, we adopt Poisson's law proposed in [21] and analyzed in [22], where the introduction of restitution impulses allows for the bouncing of mass m_1. The perfectly plastic impact law considered in[12] is obtained as a particular case when the normal restitutioncoefficient is set equal to zero. Also an infinite sequence of impacts of decreasing amplitude in finite time are possible in the model. They represent the normal transition from free-flight to contact motions.

Depending on the values of the parameters and on the state of the system, the problem can admit no solution or multiple solutions in transitions from one regime to another. This may occur in various circumstances, some of which correspond to the classical Pailevé paradox [12]. The detailed study of these situations, which entails (i) performing further investigations to check whether the multiplicity of the solution is only apparent and can be ruled out by theoretical proofs, or (ii) improving the model to eliminate its inconsistencies (by considering, for example, tangential impacts as done in [15] and experimentally verified in [24]) is out of the scopes of the present paper and is left for future works. In all examples shown in Section 6 we simply check in every transition that the obtained solution is unique and end the simulations if it is not.

Due to the complexity of the model, the dynamical response of the system is mainly investigated numerically with the combined use of time-histories, phase portraits and bifurcation diagrams. The goal of these simulations is that of determining, although in a preliminary way, the main aspects of the system non-linear dynamics, including various phenomena such as multistability, and smooth and non-smooth bifurcations.

Contrarily to what has been often done in the literature [12], [21] and [22], we allow the moving surface to have an arbitrary—and not only constant—velocity. In particular, harmonic surface motion is considered because it simulates more realistically the motion of a windscreen wiper. In this case, the most interesting part of the dynamics occurs when the velocity changes sign. This is very important also from a practical point of view, because it is just in this situation that most unwanted phenomena are observed [19].

The paper is organized as follows. We provide in Section 2 the general framework for the equations of motion of the mechanical system of Figure. 1. The three different regimes of motion and the relevant transition conditions are described in 3 and 5. Then, we present in Section 6 a series of numerical simulations aimed at highlighting the main dynamical phenomena. The paper ends with the conclusions and hints for further developments (Section 7).

EQUATIONS OF MOTION

For the plane two-mass system represented in Figure. 1, the states chosen to describe the motion are the rotation ϕ of the rigid bar which connects the masses m_1 and m_2, and the distance y of m_2 from the surface. They are collected in the vector $u=[\phi,y]^T$. The position, velocity and acceleration vectors of the masses m_1 and m_2 are

$$\mathbf{x}_{m_1} = \begin{bmatrix} l\sin\varphi \\ y - l\cos\varphi \end{bmatrix}, \quad \dot{\mathbf{x}}_{m_1} = \begin{bmatrix} l\dot\varphi\cos\varphi \\ \dot y + l\dot\varphi\sin\varphi \end{bmatrix},$$
$$\ddot{\mathbf{x}}_{m_1} = \begin{bmatrix} -l\dot\varphi^2\sin\varphi + l\ddot\varphi\cos\varphi \\ \ddot y + l\dot\varphi^2\cos\varphi + l\ddot\varphi\sin\varphi \end{bmatrix},$$

1

And

$$\mathbf{x}_{m_2} = \begin{bmatrix} 0 \\ y \end{bmatrix}, \quad \dot{\mathbf{x}}_{m_2} = \begin{bmatrix} 0 \\ \dot y \end{bmatrix}, \quad \ddot{\mathbf{x}}_{m_2} = \begin{bmatrix} 0 \\ \ddot y \end{bmatrix}.$$

2

Let $\dot s$ and $\ddot s$ denote the velocity and acceleration of the moving surface, respectively. We define displacement, velocity and acceleration of m_1 relative to the moving surface in the normal (N) and tangential (T) directions

$$g_N = y - l\cos\varphi,$$
$$\dot g_N = \dot y + l\dot\varphi\sin\varphi,$$
$$\dot g_T = l\dot\varphi\cos\varphi - \dot s,$$
$$\ddot g_N = \ddot y + l\ddot\varphi\sin\varphi + l\dot\varphi^2\cos\varphi,$$
$$\ddot g_T = l\ddot\varphi\cos\varphi - l\dot\varphi^2\sin\varphi - \ddot s.$$

3

For later use, we also define $\dot{\mathbf{g}} = [\dot g_N, \dot g_T]^T$. The kinetic energy of the system is

$$T = \tfrac{1}{2}(m_1\dot{\mathbf{x}}_{m_1}^2 + m_2\dot{\mathbf{x}}_{m_2}^2)$$
$$= \tfrac{1}{2}(m_1 + m_2)\dot y^2 + \tfrac{1}{2}m_1 l^2\dot\varphi^2 + m_1 l\dot\varphi\dot y\sin\varphi$$

4

and the potential energy is

$$V = \tfrac{1}{2}k_y(y - y_0)^2 + \tfrac{1}{2}k_\varphi(\varphi - \varphi_0)^2 + m_1 g(y - l\cos\varphi) + m_2 g y,\qquad 5$$

where g is the acceleration due to gravity, k_ϕ and k_y are the elastic coefficients of the rotational and linear springs and ϕ_0 and y_0 define the conFigureuration of the unstressed springs. The normal and tangential contact forces between the mass m_1 and the moving surface are λ_N and λ_T, and the corresponding external work is

$$P = \lambda_N(_x m_{1)2} + \lambda_T(_x m_{1)1} = \lambda_N(y - l\cos\phi) + \lambda_T l \sin\phi.\qquad 6$$

The Lagrange function is given by L=T-V+P, and the equations of motion in the general case are

$$\frac{d}{dt}\frac{\partial \mathscr{L}}{\partial \dot{\mathbf{u}}} - \frac{\partial \mathscr{L}}{\partial \mathbf{u}} = \mathbf{f}_{nc},\qquad 7$$

where $\mathbf{f}_{nc} = |-c_\varphi\dot\varphi, -c_y\dot y|^T$ are the non-conservative viscous forces of the rotational and linear dashpots represented in Figure. 1. The resulting two scalar equations, originally obtained also in [12], are

$$\mathbf{M}(\mathbf{u})\ddot{\mathbf{u}} = \mathbf{h}(\mathbf{u}, \dot{\mathbf{u}}) + \mathbf{F}(\mathbf{u})\mathbf{f},\qquad 8$$

Where

$$\mathbf{M} = \begin{bmatrix} m_1 l^2 & m_1 l \sin\varphi \\ m_1 l \sin\varphi & (m_1+m_2) \end{bmatrix}, \quad \mathbf{F} = \begin{bmatrix} l\sin\varphi & l\cos\varphi \\ 1 & 0 \end{bmatrix},$$

$$\mathbf{h} = \begin{bmatrix} -c_\varphi\dot\varphi - k_\varphi(\varphi-\varphi_0) - m_1 g l \sin\varphi \\ -c_y\dot y - k_y(y-y_0) - m_1 l\dot\varphi^2\cos\varphi - (m_1+m_2)g \end{bmatrix}, \quad \mathbf{f} = \begin{bmatrix} \lambda_N \\ \lambda_T \end{bmatrix}.\qquad 10$$

It is useful to determine explicitly the normal and tangential contact forces from (8) (note that F is always invertible apart from the uninteresting case $\phi = \pm\pi/2$): $\mathbf{f} = \mathbf{F}^{-1}\mathbf{M}\ddot{\mathbf{u}} - \mathbf{F}^{-1}\mathbf{h}$. Their explicit expressions are

$$\lambda_N = (m_1 + m_2)(\ddot y + g) + m_1 l(\ddot\varphi\sin\varphi + \dot\varphi^2\cos\varphi) + c_v\dot y + k_v(y - y_0),$$

$$\lambda_T = -m_1 l\dot\varphi^2\sin\varphi + m_1 l\ddot\varphi\cos\varphi + [c_\varphi\dot\varphi + k_\varphi(\varphi - \varphi_0)]/(l\cos\varphi)$$
$$+ \tan\varphi[-m_2(\ddot y + g) - c_v\dot y - k_v(y - y_0)].\qquad 10$$

The problem is not yet well defined: we have the four unknowns ϕ, y, λ_N and λ_T but only the two scalar equations (8). To make the problem well posed, at least in principle, it is necessary to distinguish between different regimes of motion, and to specify contact force laws and impact laws.

The first distinction is between *free-flight*, when mass m_1 is detached from the moving surface ($g_N > 0$), and contact motions, when mass m_1 is supported by the moving surface ($g_N = 0$).

In the latter case, a further distinction is required because of the friction between the moving surface and the mass m_1. *Slip* occurs when m_1 slides on the surface ($\dot{g}_T \neq 0$), and *stick* when m_1 has the same velocity of the moving surface ($\dot{g}_T = 0$). Friction is modeled according to the classical Coulomb's law, so that when $g_N = 0$ three generic cases are possible:

$$\dot{g}_T = 0 \quad \Rightarrow \quad |\lambda_T| \leqslant \mu\lambda_N \text{ (stick)},$$
$$\dot{g}_T < 0 \quad \Rightarrow \quad \lambda_T = \mu\lambda_N \text{ (negative slip)},$$
$$\dot{g}_T > 0 \quad \Rightarrow \quad \lambda_T = -\mu\lambda_N \text{ (positive slip)}. \qquad 11$$

The three different regimes of free-flight, slip and stick are analyzed separately in the following sections, together with the transition conditions from one regime to the others.

FREE-FLIGHT

Free-flight occurs when $g_N > 0$. In this case there is no contact with the moving surface, so that $\lambda_N = \lambda_T = 0$, and there are only the kinematic unknowns $u = [\phi, y]^T$ (two mechanical dof), which are determined by solving Eqs. (8) with $f = 0$. Since $\det M > 0, \forall u$, these equations reduce to

$$\ddot{u} = M^{-1}(u)h(u, \dot{u}) \qquad 12$$

and can be easily integrated numerically.

Free-flight occurs as long as $g_N > 0$, a condition that must be verified during the numerical computations. When $g_N = 0$, two situations may occur: (i) $\dot{g}_N = 0$ and (ii) $\dot{g}_N < 0$.

The first case corresponds to the so-called 'grazing' [16]. Although it is known that it plays an important role in many bifurcation scenarios, its study is out of the scopes of the present work. However, we insert a warning in the numerical code and stop the computations when a grazing is encountered.

The second case, which is the common one, is considered in the next section.

Impact

We model the impact according to Poisson's (impulse) law proposed in [21], which is summarized here and adapted to the present work. A very similar model is discussed in [10], while an energetic restitution law is considered in [17].

Although the impact is assumed to be instantaneous, only in its modeling it is customary to consider that it occurs in an infinitesimal time interval $[_{tA},_{tE}]$, in which two phases are identified: a compression, $t \in [_{tA},_{tC}]$, and an expansion, $t \in [_{tC},_{tE}]$. The compression phase is defined by the condition $\dot{g}_N < 0$, whereas the expansion phase by $\dot{g}_N > 0$. Furthermore, the position u is assumed to be continuous and constant during the impact, whereas the velocity $\dot{u}$ has a discontinuity as a consequence of the impulses generated at the impact.

Compression

Integrating Eq. (8) over the compression interval $[_{tA},_{tC}]$ we obtain

$$\mathbf{M}(\dot{\mathbf{u}}_C - \dot{\mathbf{u}}_A) = \mathbf{F}\mathbf{i}_C, \qquad 13$$

where $\mathbf{i}_C = [\Lambda_{NC}, \Lambda_{TC}]^T = [\int_{t_A}^{t_C} \lambda_N \, dt, \int_{t_A}^{t_C} \lambda_T \, dt]^T$ is the vector of the impulses of the vertical and horizontal contact forces. We notice that the integral of h in (8) vanishes because u is continuous and the time interval is infinitesimal.

Let $\mathbf{w} = [0, -\dot{s}]^T$, and rewrite (3)$_{2,3}$ in the form:

$$\dot{\mathbf{g}} = \mathbf{F}^T \dot{\mathbf{u}} + \mathbf{w}. \qquad 14$$

By evaluating (14) in $_{tA}$ and $_{tC}$, subtracting these equations, noting that F depends only on u and thus it is constant during the impact, we get

$$\dot{\mathbf{g}}_C - \dot{\mathbf{g}}_A = \mathbf{F}^T (\dot{\mathbf{u}}_C - \dot{\mathbf{u}}_A), \qquad 15$$

which, using (13), is written as function of the impulses

$$\dot{g}_C = G i_C + \dot{g}_A,$$
16

where $G = F^T M^{-1} F$. Equation (16) written explicitly reads as

$$\begin{bmatrix} \dot{g}_{NC} \\ \dot{g}_{TC} \end{bmatrix} = \begin{bmatrix} G_{11} & G_{12} \\ G_{21} & G_{22} \end{bmatrix} \begin{bmatrix} \Lambda_{NC} \\ \Lambda_{TC} \end{bmatrix} + \begin{bmatrix} \dot{g}_{NA} \\ \dot{g}_{TA} \end{bmatrix}.$$
17

The unknowns of the two scalar equations (17) are $\dot{g}_{NC}, \dot{g}_{TC}, \Lambda_{NC}$ and Λ_{TC} ($\dot{g}_{NA} < 0$ and $\dot{g}_{TA}$ are known because they are the hitting relative velocities), so that two further equations are required. They are obtained by considering the normal and the tangential behavior in compression.

The phase of compression ends, at $t = t_C$, when the normal relative velocity vanishes,

$$\dot{g}_{NC} = 0.$$

This is the third equation. Since the normal reaction λ_N is always positive, also the impulse is positive,

$$\Lambda_{NC} > 0.$$
19

As far as the behavior in the tangential direction is concerned, it is assumed that Coulomb's law (11), which is valid for forces, extends to the impulses (see [21] for more details):

$$\dot{g}_{TC} = 0 \;\Rightarrow\; |\Lambda_{TC}| \leqslant \mu \Lambda_{NC} \;\text{(stick)},$$
$$\dot{g}_{TC} < 0 \;\Rightarrow\; \Lambda_{TC} = \mu \Lambda_{NC} \;\text{(negative slip)},$$
$$\dot{g}_{TC} > 0 \;\Rightarrow\; \Lambda_{TC} = -\mu \Lambda_{NC} \;\text{(positive slip)}.$$
20

In each of the previous three cases there is an equality, which is the fourth equation, and an inequality, which has to be checked together with (19).

In the numerical simulations we explicitly consider all the three cases. If only one is possible, then the solution is unique, we take the correct values of $\dot{g}_{NC} = 0, \dot{g}_{TC}, \Lambda_{NC}$ and Λ_{TC}, and proceed further. If, on the contrary, none or more than one is possible, we encounter a paradox, and, according to what has been said in the Introduction, we terminate the computation.

Finally, we note that the behavior in the tangential direction is dissipative, since $\Lambda_{TC} \cdot \dot{g}_{TC} \leqslant 0$ in any case.

Expansion

Repeating the same reasoning of the compression phase in the infinitesimal interval $[t_C, t_E]$ we obtain (G is the same of the compression phase):

$$\dot{g}_E = Gi_E + \dot{g}_C.$$

21

Now $i_E = [\Lambda_{NE}, \Lambda_{TE}]^T = [\int_{t_C}^{t_E} \lambda_N dt, \int_{t_C}^{t_E} \lambda_T dt]^T$ is the vector of the impulses in the expansion phase.

Again, two further equations are required to determine the four unknowns $\dot{g}_{NE}, \dot{g}_{TE}, \Lambda_{NE}$ and Λ_{TE} ($\dot{g}_{NC} = 0$ and $\dot{g}_{TC}$ have been computed in the compression phase).

The behavior in the normal direction is governed by Poisson's law, according to which the elastic deformation in compression releases an expansion impulse which is proportional to the impulse at compression:

$$\Lambda_{NE} = \varepsilon_N \Lambda_{NC}.$$

22

This is the third equation. $\varepsilon_N \geqslant 0$ is the restitution coefficient, and it is a constitutive parameter of the model which should be determined experimentally.

The relative normal velocity at the end of the expansion must be non-negative, $\dot{g}_{NE} \geqslant 0$. Indeed, it has been shown in [11] that the mass m_1 always rebound if $\varepsilon_N > 0$, namely

$$\dot{g}_{NE} > 0.$$

23

The behavior in the tangential direction is similar to that in the compression phase, and it is given by (20) by simply changing the subscript C with E.

Again, in each of the three cases there is an equality, which is the fourth equation, and an inequality, which has to be checked together with (23).

In the numerical simulations we consider all the three cases. If only one is possible, then the solution is unique, we get the corresponding $\dot{g}_{NE}, \dot{g}_{TE}, \Lambda_{NE}$ and Λ_{TE}, and proceed further. If, on the contrary, none or more than one is possible, we encounter a paradox and stop the computation.

Remark.

In [21] a more involved model is considered, which is aimed at describing also the 'superball' effect. See[21] and [11] for more details. In [15] and [24] a tangential impact is also considered.

Summarizing, we enter the impact phase knowing the impact position $_\phi A$ and $_yA = \operatorname{lcos}(_\phi A)$, and the velocities $\dot\varphi_A$ and $\dot y_A$. With these latter we compute $\dot g_{NA}$ and $\dot g_{TA}$ by $(3)_{2,3}$ and then, following the previous developments, $\dot g_{NE}$ and $\dot g_{TE}$. Using again $(3)_{2,3}$ we get the initial velocities $\dot\varphi_E$ and $\dot y_E$ for the subsequent motion, while the continuity of the displacements gives $_\phi E = _\phi A$ and $_yE = _yA$.

Note that all these computations are algebraic, and that the (infinitesimal) length of the time interval $[_{tA}, _{tE}]$ does not appear explicitly. This confirms that the impact is actually considered instantaneous in this model.

From Free-Flight to Contact Regimes
Infinitely Many Impacts

As previously said, it has been shown in [11] that $\dot g_{NE} > 0$ provided that $_\varepsilon N > 0$. This means that the mass$_{m1}$ always bounces, and there is no *generic direct* transition from free-flight to contact regimes. This transition in general occurs *indirectly* . In fact, in certain situations an infinite sequence of impacts occur at times $_{ti}$, $i \in \mathbb{N}$, and the impact instants accumulate on a certain $\hat t$, $\lim_{i \to \infty} t_i = \hat t$, while the normal impulses tend to zero, $\lim_{i \to \infty} A_{NC,i} = 0$, as well as the normal relative velocity, $\lim_{i \to \infty} \dot g_{NE,i} = 0$.

In the numerical simulations it is not easy to detect the accumulation point $\hat t$. We conventionally end the process when the time distance between two consecutive impacts, $|_{ti+1} - _{ti}|$, is smaller than a given, very small, tolerance.

At the end of this process, which has been studied in detail in [2] and [5] where it has been named 'chattering', and in [18] where it has been named 'complete chattering', the motion goes to a contact regime. To decide whether stick or slip is the subsequent motion, we use the exit condition for the impact as the initial conditions for both stick and slip, and immediately check their feasibility (practically, for slip we check if $_{\lambda N}$ given by (31) is positive, and for stick we check if $_{\lambda T}$ and $_{\lambda N}$ computed with the formulas reported in Section 5 satisfy $|_{\lambda T}| < \mu_{\lambda N}$). If only one is admissible, the motion proceed with that phase, otherwise a paradox is encountered and, as usual, we stop the numerical simulations.

Completely Inelastic Impact

A different transition from free-flight to contact regimes is obtained in the case of a perfectly plastic impact. This is obtained by setting $\varepsilon_N = 0$, i.e., by assuming that there is no expansion impulse or, alternatively, that all the normal impact energy is dissipated.

In this case the expansion phase disappears, $_tE \equiv {}_tC$, and the solution is that of the compression phase

$$\dot{g}_{NE} = \dot{g}_{NC} = 0, \quad \dot{g}_{TE} = \dot{g}_{TC}. \tag{24}$$

Remark.

The perfectly plastic impact can be seen as the limit of Poisson's impact law described in the previous sections as $_tC \rightarrow {}_tE$ (see [11] for more details). A perfectly plastic impact is considered in [12].

Contrary to the previous case, here the transition to contact regimes is instantaneous, and does not require any special care in the numerical simulations. Similar to the previous case, instead, the contact regime following the impact is chosen by checking the feasibility of both slip and stick motions, continuing the computation if only one is possible and stopping it if none or both are possible.

Remark.

It is useful to determine the manifold which separates the stick and slip post-impact regions in the phase space. This manifold can be detected by assuming that the stick post-impact condition ($\dot{g}_{TE} = 0$, namely $\dot{g}_{TC} = 0$) and the slip post-impact condition ($_\Lambda TE = \mu_\Lambda NE$, namely $_\Lambda TC = \mu_\Lambda NC$) occur simultaneously. With these conditions, from (17) we obtain that $\Lambda_{NC} = -\dot{g}_{NA}/(G_{11} + \mu G_{21})$ and that the equation

$$H(\varphi_A)\dot{g}_{NA}(\varphi_A, \dot{\varphi}_A, \dot{y}_A) = \dot{g}_{TA}(\varphi_A, \dot{\varphi}_A) \quad \text{with} \quad H = \frac{G_{21} + \mu G_{22}}{G_{11} + \mu G_{12}}, \tag{25}$$

must be satisfied by the impact entering variables $_\phi A$, $\dot{\varphi}_A$ and $\dot{y}_A$. Eq. (25) is the analytical expression of the searched boundary manifold. Note that in the sub-space $(\phi, \dot{\varphi})$, it is a curve parametrized by $\dot{y}_A$.

SLIP

Slip is the first of the two contact regimes to be analyzed. Both are characterized by $_gN = 0$ and $\dot{g}_N = 0$, and as far as contact is maintained, also by $\ddot{g}_N = 0$. From these expressions and (3) we obtain

$$y = l \cos \varphi, \quad \dot{y} = -l \dot{\varphi} \sin \varphi, \quad \ddot{y} = -l \dot{\varphi}^2 \cos \varphi - l \ddot{\varphi} \sin \varphi, \qquad 26$$

which show how, in these cases, y is no longer an independent variable.
In the case of slip, ϕ is the unique independent variable (1 dof). To determine the governing equation, we note that if m_1 slides on the moving surface, then $\dot{g}_T \neq 0$, and, from the Coulomb law $(11)_{2,3}$, we have that:

$$\lambda_T = \mu \lambda_N \ (\text{if } \dot{g}_T < 0) \ \text{or} \ \lambda_T = -\mu \lambda_N \ (\text{if } \dot{g}_T > 0). \qquad 27$$

We write all equations in the present section by supposing $\lambda_T = \mu \lambda_N$; the other case is obtained by changing μ with $-\mu$ (alternatively, we can consider $\lambda_T = -\text{sign}(\dot{g}_T)\mu\lambda_N$). By inserting expression (10)in the expression $\lambda_T = \mu \lambda_N$, and using (26), we get (see [12])

$$m(\varphi)\ddot{\phi} + v(\dot{\phi}, \varphi) = 0, \qquad 28$$

Where

$$m(\varphi) = (m_1 \cos^2 \varphi + m_2 \sin^2 \varphi + \mu m_2 \sin \varphi \cos \varphi)l^2,$$
$$v(\dot{\phi}, \varphi) = a_1(\varphi)\dot{\phi}^2 + a_2(\varphi)\dot{\phi} + a_3(\varphi), \qquad 29$$

And

$a_1(\phi)=[(m_2-m_1)\sin\phi\cos\phi+\mu m_2\cos^2\phi]l^2, a_2(\phi)=\mu c_v l^2\cos\phi\sin\phi+c_v l^2\sin^2\phi+c_\phi, a_3(\phi)=k_\phi(\phi-\phi_0)-k_v l^2\sin\phi\cos\phi-\mu k_v l^2\cos^2\phi+k_v y_0 l\sin\phi+\mu k_v y_0 l\cos\phi-(m_1+m_2)(\sin\phi+\mu\cos\phi)gl+m_1 gl\sin\phi.$

$$30$$

Eq. (28) is integrated numerically. Once the solution $\phi(t)$ is determined, the normal reaction λ_N is evaluated by substituting (26) and (28) in $(10)_1$:

$$\lambda_N = m_2 l \left(\frac{v(\dot{\phi}, \varphi)}{m(\varphi)} \sin \varphi - \dot{\phi}^2 \cos \varphi \right) - c_v l \dot{\phi} \sin \varphi$$
$$+ k_v l \cos \varphi - k_v y_0 + (m_1 + m_2)g. \qquad 31$$

Slip occurs if $m(\phi)>0$, $\dot{g}_T \neq 0$ and $\lambda_N>0$, three conditions that must be verified during the numerical computations. When $m(\phi)\leqslant 0$, the Painlevé paradox occurs (Section 4.2); when $\dot{g}_T = 0$ we have a transition to stick or to inverted slip (Section 4.3); when λ_N vanishes we have a transition to free-flight (Section 4.4). Before analyzing these cases, however, we discuss the equilibrium points in the slip regime.

Equilibrium Points and Their Stability

Equilibrium positions $\varphi = \tilde{\varphi}$ and $y = l\cos\tilde{\varphi}$ of the system of Figure. 1 are possible if the surface velocity does not change sign, e.g., if it is constant. They experience only the slip regime, so that, from (28), they are the solutions to $a_3(\tilde{\varphi}) = 0$.

The determination of the stability of these solutions is a difficult problem. In fact, they are (equilibrium) solutions to the whole 2 dof system, so that the computation of the eigenvalues of the whole 4×4 Jacobian matrix is required. Moreover, the problem is even more complex, because we are dealing with a non-smooth unilaterally constrained motion, so that classical analysis may fail [14].

As a consequence of the previous observation, and because it is sufficient for the purposes of the present work, we study only the stability *on* the constrained manifold $_gN=0$, i.e., we investigate only the stability of the equilibrium of Eq. (28), neglecting the out-of-manifold dimensions. Thus, we only obtain necessary conditions for stability (or sufficient conditions for instability). Fortunately, we will show with some numerical simulations (Section 6.1) that, in some cases, this analysis captures exactly the bifurcation point where stability is lost.

The Jacobian matrix of system (28) at equilibrium is given by (see [11] for more details)

$$\mathbf{J} = \frac{-1}{m(\tilde{\varphi})}\begin{bmatrix} 0 & -m(\tilde{\varphi}) \\ a_3'(\tilde{\varphi}) & a_2(\tilde{\varphi}) \end{bmatrix}.$$

32

Let $\lambda_1(\tilde{\varphi})$ and $\lambda_2(\tilde{\varphi})$ be the eigenvalues of (32). The equilibrium is stable if $\mathrm{Re}(\lambda_1)<0$ and $\mathrm{Re}(\lambda_2)<0$.

The equilibrium reaches a saddle-node bifurcation if one of the eigenvalues is equal to zero. This occurs when $a_3'(\tilde{\varphi})=0$, i.e., when $a_3(\phi)$ has a double zero. The bifurcation is instead a Hopf when $\mathrm{Re}(\lambda_i)=0, \mathrm{Im}(\lambda_i)\neq0, i=1,2$. This occurs when $a_2(\tilde{\varphi})=0$ and $a_3'(\tilde{\varphi})>0$. Note that if $c_\varphi > (c_v l^2/2)(\sqrt{1+\mu^2}-1)$ we have that $a_2(\phi)$ is always different from zero, so that Hopf bifurcation is not possible.

PainlevéParadox

In (28), $m(\phi)$ vanishes when

$$\varphi = \arctan\left[\frac{\mu}{2}\left(-1 \pm \sqrt{1 - \frac{4}{\mu^2}\frac{m_1}{m_2}}\right)\right],$$

33

which is a real number only if [12]

$$\mu > \mu_{crit} = 2\sqrt{\frac{m_1}{m_2}}.$$

34

If $\mu < \mu_{crit}$ the mass term $m(\phi)$ never vanishes, whereas for μ above μ_{crit} it is possible that the term $m(\phi)$ vanishes. When $m(\phi)=0$, two cases are possible:

(i) $v(\dot{\varphi}, \varphi) \neq 0$; then $\ddot{\varphi} \to \infty$ and the solution no longer exist;

(ii) $v(\dot{\varphi}, \varphi) = 0$, and the solution is not unique in principle.

This is a manifestation of the Painlevé paradox, which can be better studied by reformulating the problem in the form of a linear complementary problem (LCP). This has been previously investigated in [12] and [21], and their main results are summarized in the next subsection.

The Linear Complementary Problem

Let us consider expression $(3)_4$ of $\ddot{g}_N$. If we (i) evaluate $\ddot{\varphi}$ and $\ddot{y}$ from the equation of motion (8), (ii) use the condition of slip $\lambda_T = \mu \lambda_N$ (we suppose positive sliding) and (iii) use the conditions of contact $g_N = \dot{g}_N = 0$ (i.e., $y = l\cos\phi$ and $\dot{y} = -l\dot{\varphi}\sin\varphi$), then we have [12]

$$\ddot{g}_N = A(\varphi)\lambda_N + b(\dot{\varphi}, \varphi),$$

35

Where

$$A(\varphi) = \frac{1}{(m_1\cos^2\varphi + m_2)}\frac{m_2}{m_1}\left(\frac{m_1}{m_2}\cos^2\varphi + \sin^2\varphi + \mu\sin\varphi\cos\varphi\right)$$

$$b(\dot{\varphi}, \varphi) = \frac{1}{l(m_1\cos^2\varphi + m_2)}\left\{m_2 l^2\cos\varphi\dot{\varphi}^2\right.$$

$$+ \left[-c_\varphi\frac{m_2}{m_1}\sin\varphi + c_v l^2\sin\varphi\cos^2\varphi\right]\dot{\varphi}$$

$$\left.+ \left[-k_v l^2\cos^3\varphi - k_\varphi\frac{m_2}{m_1}(\varphi - \varphi_0)\sin\varphi + lk_v y_0\cos^2\varphi\right]\right\} - g.$$

36

In the case of negative sliding, $\lambda_T = -\mu\lambda_N$ and in (36)μ has to be replaced with $-\mu$.

When the mass is in contact with the moving surface, the force λ_N can be greater than zero and $\ddot{g}_N$ must be zero, or $\ddot{g}_N$ can be greater than zero and λ_N must be zero. We thus obtain the LCP (35) and

$$\ddot{g}_N \geq 0, \quad \lambda_N \geq 0, \quad \ddot{g}_N \cdot \lambda_N = 0. \qquad 37$$

The following cases are possible:
• A>0 (one solution)
  (1) $b>0 \rightarrow \lambda_N=0, \ddot{g}_N = b$ (free-flight);
  (2) $b<0 \rightarrow \lambda_N=-b/A, \ddot{g}_N = 0$ (slip);
• A<0 (two solutions or no solutions)
  (1) $b>0 \rightarrow \lambda_N=0, \ddot{g}_N = b$ (free-flight), or $\lambda_N=-b/A, \ddot{g}_N = 0$ (slip)
  (2) $b<0 \rightarrow$ no solutions.

We note that $A(\phi)=m(\phi)/[m_1(m_2+m_1\cos^2\phi)]$, so that $A(\phi)=0$ if and only if $m(\phi)=0$. Moreover, when the Painlevé paradox occurs, $A(\phi)<0$ [21], also the mass-like term $m(\phi)$ becomes negative.

From Slip to Stick or To Inverted Slip

The motion passes from slip to stick or to inverted slip when $\dot{g}_T = 0$. This occurs at a certain $\hat{\varphi}$, and the associated velocity and acceleration are, respectively,

$$\hat{\dot{\varphi}} = \frac{\dot{s}}{l\cos(\hat{\varphi})}, \qquad \hat{\ddot{\varphi}} = -\frac{v(\hat{\varphi}, \hat{\dot{\varphi}})}{m(\hat{\varphi})}. \qquad 38$$

They depend on the transition position $\hat{\varphi}$ only.

To decide the next regime, we check independently the admissibility of impending stick or inverted slip.

Stick after Slip

To verify if stick follows slip, we compute the tangential λ_T^+ and the normal λ_N^+ contact forces at the beginning of the hypothetical subsequent stick regime. They are given by inserting expressions (26) of y, $\dot{y}$ and $\ddot{y}$ in (10), and by using in the resulting formulas $\varphi = \hat{\varphi}$,

$$\dot{\varphi} = \frac{\dot{s}}{l\cos\hat{\varphi}}, \quad \ddot{\varphi} = \frac{\ddot{s}}{l\cos\hat{\varphi}} + \frac{(\dot{s})^2\sin\hat{\varphi}}{l^2\cos^3\hat{\varphi}}$$

39

(these are the equations of the stick regime, which will be obtained in the next section).

Inverted Slip

To check whether inverted slip might occur at $\dot{g}_T = 0$, we compute the tangential relative acceleration at the end of the finishing slip regime

$$\ddot{g}_T^- = l\hat{\ddot{\varphi}}\cos\hat{\varphi} - l\hat{\dot{\varphi}}^2\sin\hat{\varphi} - \ddot{s},$$

40

where $\hat{\dot{\varphi}}$ and $\hat{\ddot{\varphi}}$ are given by (38), so that (40) is a function of $\hat{\varphi}$ only.

Then, we compute the tangential relative acceleration at the beginning of the hypothetical next slip regime, $\ddot{g}_T^+$. It is given by (40) by simply changing the sign of μ in the expression $\nu\langle\hat{\varphi}, \hat{\varphi}\rangle$ appearing in (38)$_2$.

Inverted slip is possible if the relative velocity $\dot{g}_T$ continuously changes sign across the point where $\dot{g}_T = 0$. This occurs when $\ddot{g}_T^+$ and $\ddot{g}_T^-$ have the same sign, i.e., $\ddot{g}_T^+ \cdot \ddot{g}_T^- > 0$.

Transition at the End of Slip

Now that we have (independently) computed $\ddot{g}_T^+$, $\ddot{g}_T^-$, λ_T^+ and λ_T^- we are in position to determine the transition at the end of the slip regime. Only four generic cases are possible:

(i) $|\lambda_T^+| \leqslant \mu\lambda_N^+$ and $\ddot{g}_T^+ \cdot \ddot{g}_T^- < 0 \Rightarrow$ we switch to stick, which is the only possible subsequent motion;

(ii) $|\lambda_T^+| \geqslant \mu\lambda_N^+$ and $\ddot{g}_T^+ \cdot \ddot{g}_T^- > 0 \Rightarrow$ we switch to inverted slip, which is the unique possibility;

(iii) $|\lambda_T^+| > \mu\lambda_N^+$ and $\ddot{g}_T^+ \cdot \ddot{g}_T^- < 0 \Rightarrow$ no solution exists.

(iv) $|\lambda_T^+| \leqslant \mu\lambda_N^+$ and $\ddot{g}_T^+ \cdot \ddot{g}_T^- > 0 \Rightarrow$ two solutions exist;

Note that cases (iii) and (iv) highlight the possible presence of another paradox. In these cases the numerical simulations are interrupted, as well as in the non-generic situation $\ddot{g}_T^+ \cdot \ddot{g}_T^- = 0$ which is similar to grazing.

From Slip to Free-Flight

The motion moves from slip to free-flight when $\lambda_N=0$, which, by (35), is equivalent to b=0 ($\ddot{g}_N = 0$ because we are in a slip regime, although at the end of it). Consequently, in the phase space ($\phi,\dot{\phi}$), the curve implicitly defined by $b(\dot{\phi}, \phi) = 0$ is the locus of points of slip to free-flight transition.

Note that during the transition the function b behaves as follows: b<0 (slip) $\Rightarrow$ b=0 (transition) $\Rightarrow$ b>0 (free-flight).

Let $\hat{\phi}$ and $\hat{\dot{\phi}}$ be the angle and the velocity at the instant of time for which $\lambda_N=0$. The initial conditions for the subsequent free-flight regime are: $\phi = \hat{\phi}$, $\dot{\phi} = \hat{\dot{\phi}}$, $y = l \cos \hat{\phi}$ and $\dot{y} = -l\hat{\dot{\phi}} \sin \hat{\phi}$.

STICK

In the stick case the mass m_1 adheres to the moving surface, so that the motion has no dof. In addition to (26), we also have $\dot{g}_T = \ddot{g}_T = 0$, which yields

$$\dot{\phi} = \frac{\dot{s}}{l \cos \phi}, \quad \ddot{\phi} = \frac{\ddot{s}}{l \cos \phi} + \frac{(\dot{s})^2 \sin \phi}{l^2 \cos^3 \phi}.$$

$$41$$

Integrating $\dot{g}_T = 0$ we obtain

$$\phi(t) = \arcsin \left[\sin \hat{\phi} + \frac{1}{l} \int_{\hat{t}}^{t} \dot{s}(\tau)d\tau \right],$$

$$42$$

which gives the angle $\phi(t)$ as a function of the driving velocity. In (42) $\hat{\phi}$ is the angle and $\hat{t}$ the time at the beginning of the considered stick regime.

If we substitute (41) and (42) in (26), and the result in (10), we obtain the reactive forces λ_N and λ_T as a function of $\dot{s}$ and $\ddot{s}$.

Note that in the space $(\varphi, \dot{\varphi})$ the stick motions belong to the curve defined by $(41)_1$, which is explicitly independent on time only in the case of constant surface velocity.

Stick occurs as far as $|\lambda_T| < \mu\lambda_N$ and $\lambda_N > 0$. When $|\lambda_T| = \mu\lambda_N$ we have a transition to slip (Section 5.1); when $\lambda_N = 0$ we have a transition to free-flight (Section 5.2).

From Stick to Slip

A stick phase passes to slip when $\lambda_T = \mu\lambda_N$ (negative slip, $\dot{g}_T < 0$) or $\lambda_T = -\mu\lambda_N$, (positive slip, $\dot{g}_T > 0$).

Of course, we have to check that incipient slip is possible. λ_N remained positive during all the stick phase (otherwise there would be transition to free-flight), so it is positive also at the transition and just after. The relative tangential velocity $\dot{g}_T$ is zero up to the transition and then continuously will become different from zero. The only meaningful check is then on the sign of $m(\phi)$ (and thus on the sign of $A(\phi)$), which may turn to a negative value during the stick phase.

If $m(\phi)$ (and $A(\phi)$) is positive, according to the contents of Section 4.2.1 the solution is unique and we proceed further. If $m(\phi) < 0$ (and $A(\phi) < 0$) the Painlevé paradox occurs, and we have to consider the sign of b. If b is negative there is no solution, and the computations stop. If, on the other hand, $b > 0$ we can have slip or free-flight after transition (see Section 4.2.1).

Since we do not accept negative values of the mass-like term $m(\phi)$, in this case we proceed to free-flight. In this transition λ_N jumps instantaneously from a positive value to zero. This choice is supported by the results of [12, Lemma 5.3], where it is proved, for a simplified frictional impact oscillator, that the unique transition is from stick to free-flight.

From Stick to Free-Flight

The transition from stick to free-flight apparently does not differ from the slip to free-flight transition which has been studied in Section 4.4. A difference instead occurs. In fact, when $\lambda_N = 0$ also $\lambda_T = 0$, so that both conditions $\lambda_N = 0$ and $|\lambda_T| = \mu\lambda_N$ are satisfied simultaneously. Thus, in principle, both the transitions to free-flight and to slip are possible. As usual, we proceed if only one is possible, and stop if none or both are feasible.

It is important to remark that, in some sense, this case constitutes a pathological situation. In fact, it requires that, during the stick phase, the point (λ_N, λ_T) in the plane of the contact forces reaches the vertex of the friction cone with no previous intersection with the cone itself (otherwise a transition to slip will happen earlier), which is not a generic situation. To support this conclusion, we observe that in the numerical simulations of Section 6 we never encounter a stick to free-flight transition.

NUMERICAL SIMULATIONS

In this section two series of numerical simulations are reported.

In the first we assume a constant surface velocity and we analyze the influence on the dynamics of the frictional coefficient μ and of the angle of the unstressed conFigureuration ϕ_0 on the dynamics. Different values of the restitution coefficient ε_N and of the ratio m_1/k_ϕ are considered.

In the second we consider a sinusoidal surface velocity. Preliminary tests are performed to investigate the hopping phenomenon, which usually takes place when ϕ and $\dot{s}$ have opposite signs, i.e. when the motion of the surface compresses the vertical spring. This phenomenon is unwanted in many practical applications. If, for example, we consider the motion of windscreen wipers, the bounces of the blade on the glass produce undesired noise and marks which reduce the visibility. In the case of robotic fingers, the hopping reduces the displacement precision and stresses the system by means of dangerous impulsive forces. We analyze the influence of the friction and of the unstressed springscon Figureurations on the hopping motion.

Constant Surface Velocity

We set $l = 1\,\mathrm{m}$, $m_2 = 1\,\mathrm{kg}$, $k_y = 100\,\mathrm{N/m}$, $c_y = 10\,\mathrm{N/(ms)}$, $c_\phi = 0$, $y_0 = 1\,\mathrm{m}$ and $\dot{s} = -1\,\mathrm{m/s}$, and consider both an elasto-plastic impact law, assuming $\varepsilon_N = 0.2$, and a perfectly plastic law as in [12], setting $\varepsilon_N = 0$. The friction coefficient μ and the angle of the unstressed conFigureuration ϕ_0 of the rotational spring play the role of governing parameters. Two simulations are performed with different values of the ratio m_1/k_ϕ.

Simulation 1

We consider the values $m_1 = 0.1\,\mathrm{kg}$ and $k_\varphi = 100\,\mathrm{Nm}$. For $\varepsilon_N = 0$ (perfectly plastic impact), the chosen parameters coincide with those used in the simulations of [12], which are used as a benchmark to check the accuracy of our numerical code. Then, they are compared with the results obtained with $\varepsilon_N=0.2$ in order to have some hints on the differences induced by different impact laws.

Bifurcation Diagrams for Varying M
We fix $\phi_0=\pi/8$. The influence of the friction coefficient μ on the motion is described by the bifurcation diagrams of Figure. 2. The maximum angular velocity $\dot\varphi$ of the motion is plotted versus μ.

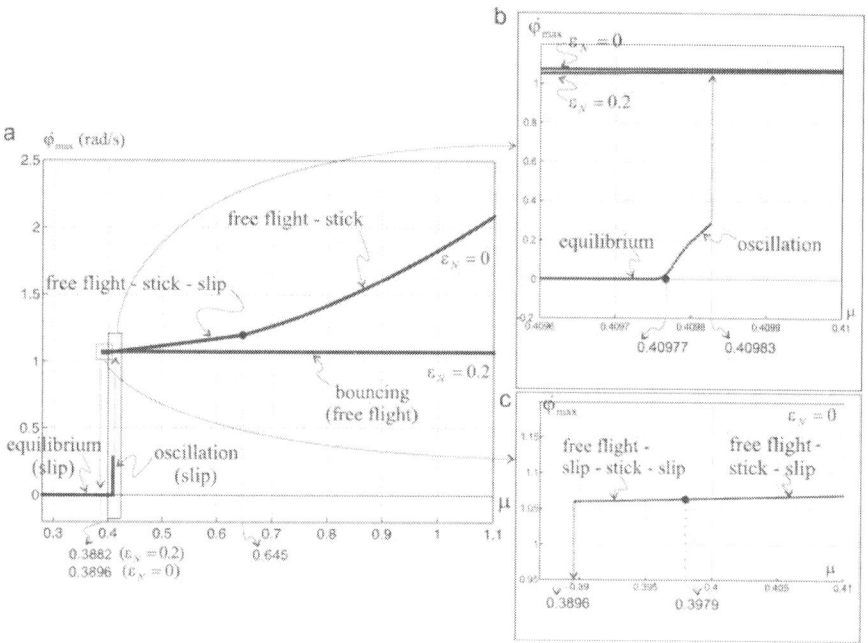

Figure 2. Simulation 1. Bifurcation diagrams for varying μ.

For small values of μ we observe only an equilibrium branch, which corresponds to that analytically computed in Section 4.1. As it is in a slip mode, no impacts are involved and thus the diagrams for $\varepsilon_N=0$ and $\varepsilon_N=0.2$ coincide. Furthermore, the equilibrium branch is horizontal, being $\dot\varphi$ constantly equal to zero, but the equilibrium conFigureurations actually depend on μ: ϕ increases as μ increases, according to the equilibrium equation $a_3(\phi)=0$, see Section 4.1.

By increasing μ, the equilibrium undergoes anHopf bifurcation at μ=0.40977, after which a periodic oscillation in slip mode appears. This bifurcation occurs within the slip mode, so that the bifurcation point coincides with that analytically computed in Section 4.1. Furthermore, as the system remains in a slip mode, the diagrams for $_{\varepsilon N}$=0 and 0.2 still coincide. The equilibrium point $(\varphi = 0.392, \dot{\varphi} = 0)$ for μ=0.3 and the periodic oscillation for μ=0.4098 are reported in Figure. 3.

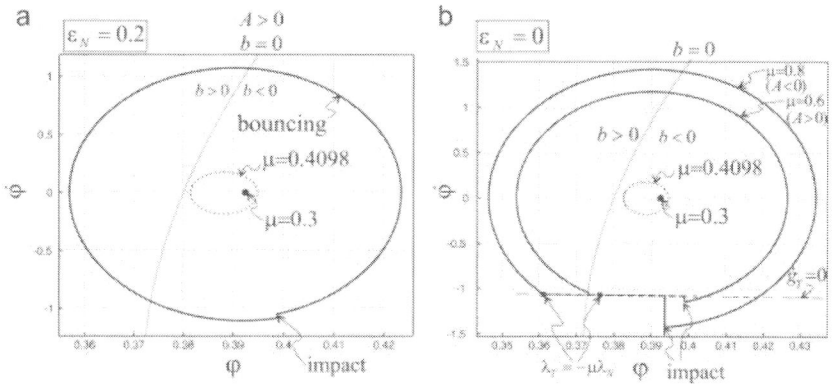

Figure 3. Simulation 1. Orbits in the φ–$\dot{\varphi}$ phase plane. Free-flight ≡ solid line; slip ≡ dotted line; stick ≡ dashed line.

The slip periodic branch exists only in a narrow range of values of μ. In fact, at μ=0.40983 it suddenly disappears and the dynamics jump on another branch of solutions, which exists also for smaller values of μ. This branch involves free-flight and impacts, and thus the path for $_{\varepsilon N}$=0 is different from that for $_{\varepsilon N}$=0.2. The two branches differ even for the starting point, which corresponds to μ≃0.3882 for $_{\varepsilon N}$=0.2 and toμ≃0.3896 for $_{\varepsilon N}$=0.

It is worth to note how the transition from contact motions to detached motions occurs through a hysteretic loop (see Figure. 2(b)), which is defined by the region of coexistence of the two solutions.

In the case $_{\varepsilon N}$=0.2, the detached motion consists of a sequence of periodic bounces of $_{m1}$ on the moving surface. Its orbit is denoted as 'bouncing' in Figure. 3(a), while a time history is reported in Figure. 4(a). It has a single jump of $\dot{\varphi}$, corresponding to the impact. The periodic motion does not depend on μ since the friction enters only at the impact instants (a numerable set of instants) and does not affect significantly the dynamics in this case. As a result, the periodic branch is horizontal.

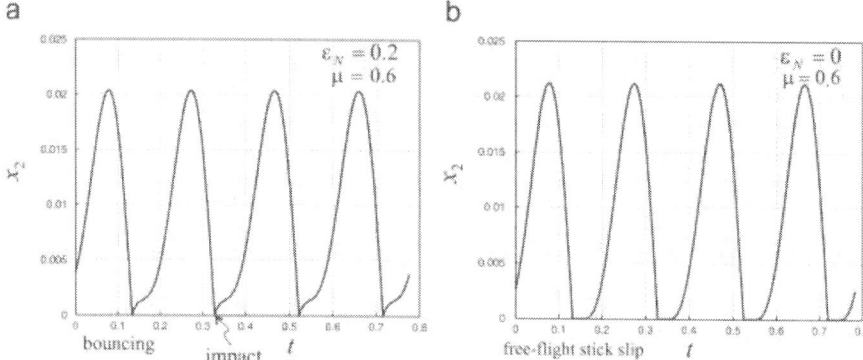

Figure 4. Simulation 1. Vertical displacement x_2 of m_1 as function of time t in the periodic motions obtained when $\mu = 0.6$.

In the case $\varepsilon_N = 0$ the motion consists of a sequence of free flight–stick–slip phases which repeat periodically, this occurring in the range $0.3979 < \mu < 0.6450$. The phase portrait for $\mu = 0.6$ is drawn in Figure. 3(b), and the corresponding time history is reported in Figure. 4(b). After the free-flight phase, at the impact the orbit jumps to the curve $\dot{g}_T = 0$ (Eq. (41)$_1$), and the motion continues in stick mode. There is a transition to slip mode when the stick curve intersects the friction cone ($\lambda_T = -\mu \lambda_N$), and then a transition from slip to free-flight occurs when the orbit intersects the curve b=0 (b is defined in (36)$_2$), as described in Section 4.2.1.

Near the bifurcation point $\mu = 0.3896$ (see Figure. 2(c) for an enlargement of the corresponding bifurcation branch) the motion has the following behavior. At $\mu = 0.3979$, the periodic motion passes from free flight–stick–slip ($\mu > 0.3979$, Figure. 3(b)) to free flight–slip–stick–slip ($\mu < 0.3979$, Figure. 5(b)), i.e., the post-impact phase passes from stick to slip. The phase portrait at this transition is represented in Figure. 5(a), where we note that the impacting values ϕ_A, $\dot{\phi}_A$ and $\dot{y}_A$ satisfy Eq. (25)$H\dot{g}_{NA} = \dot{g}_{TA}$.

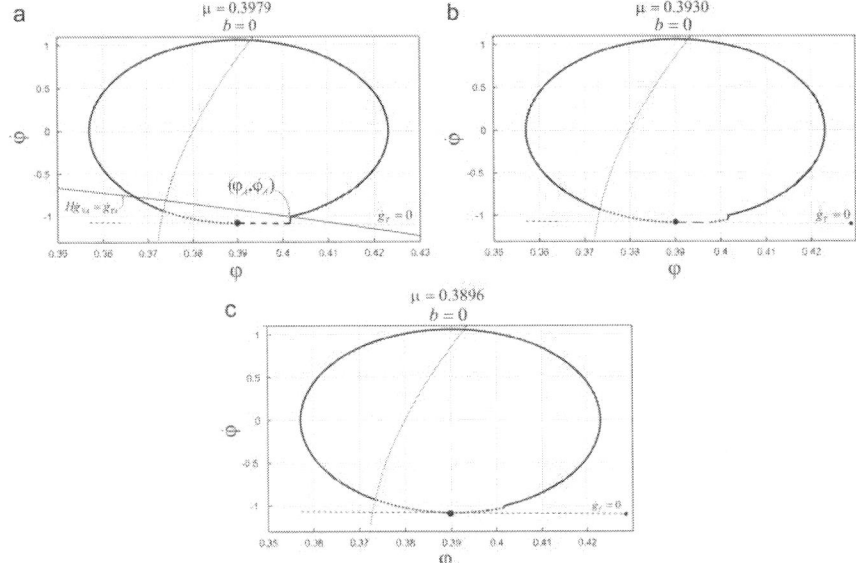

Figure 5. Simulation 1. Orbits in the $\varphi - \dot\varphi$ phase plane. Free-flight ≡ solid line; slip ≡ dotted line;stick ≡ dashed line.

As µ decreases in the range $0.3896 < \mu < 0.3979$, the stick phase reduces and the post-impact slip phase increases. At the bifurcation point $\mu \simeq 0.3896$, the stick phase disappears and the two slip parts of the orbit join at the point of intersection between the friction cone and the curve $\dot{g}_T = 0$ (see Figure. 5(c)). This bifurcation looks to be quite complex, and it is still subject to further investigation, but, in any case, after that the motion jumps to the equilibrium branch $\dot\varphi = 0$.

When $\mu > 0.645$ the periodic motion loses the slip phase (see the phase portrait for $\mu = 0.8$ in Figure. 3(b)). In this case $A < 0$ (A is defined in $(36)_1$), and the Painlevé paradox occurs at the end of the stick phase, when $\lambda_T = -\mu\lambda_N$. Since $b > 0$ at this point, both free-flight and slip would be possible (see Section 4.2.1), but the negative value of the mass-like term $m(\phi)$ rules out the slip mode and the motion goes in free-flight. However, at the transition instant from stick to free-flight, the normal reaction λ_N jumps from a positive value to zero. This discontinuity has no mechanical justification, and likely disappears if more sophisticated impact laws are considered.

The loss of the slip phase at $\mu = 0.645$ is evidenced by a non-smooth point in the bifurcation diagram ofFigure. 2(a).

A comparison of the plastic and of the elasto-plastic impact cases for fixed values of µ is reported in Figure. 3 and Figure. 4. In particular, comparing the time histories in Figure. 4, we notice that the motions for $\varepsilon N=0$ and $\varepsilon N=0.2$ have practically the same period.

Bifurcation Diagrams for Varying ϕ_0

We now fix $\mu=0.4$ and vary ϕ_0 in the interval $(0,\pi/5)$. The bifurcation diagrams are represented in Figure. 6. The solid and dashed line curves correspond to the case $\varepsilon N=0.2$ and $\varepsilon N=0$, respectively. For $\phi_0<0.223$, the mass m_1 never detaches from the belt. An equilibrium branch in slip exists up to $\phi_0=0.044$, and a stick–slip periodic branch is found in the interval $0.044<\phi_0<0.223$. Within the stick–slip interval, as ϕ_0 increases, the stick phase increases (see the phase portraits of Figure. 7). At $\phi_0=0.223$ the orbit becomes tangent to the curve b=0 and the free-flight phase enters the motion. For $\phi_0>0.223$ the diagrams corresponding to $\varepsilon N=0$ and $\varepsilon N=0.2$ do not coincide, despite their closeness up to $\phi_0\approx0.36$. We consider separately the two cases.

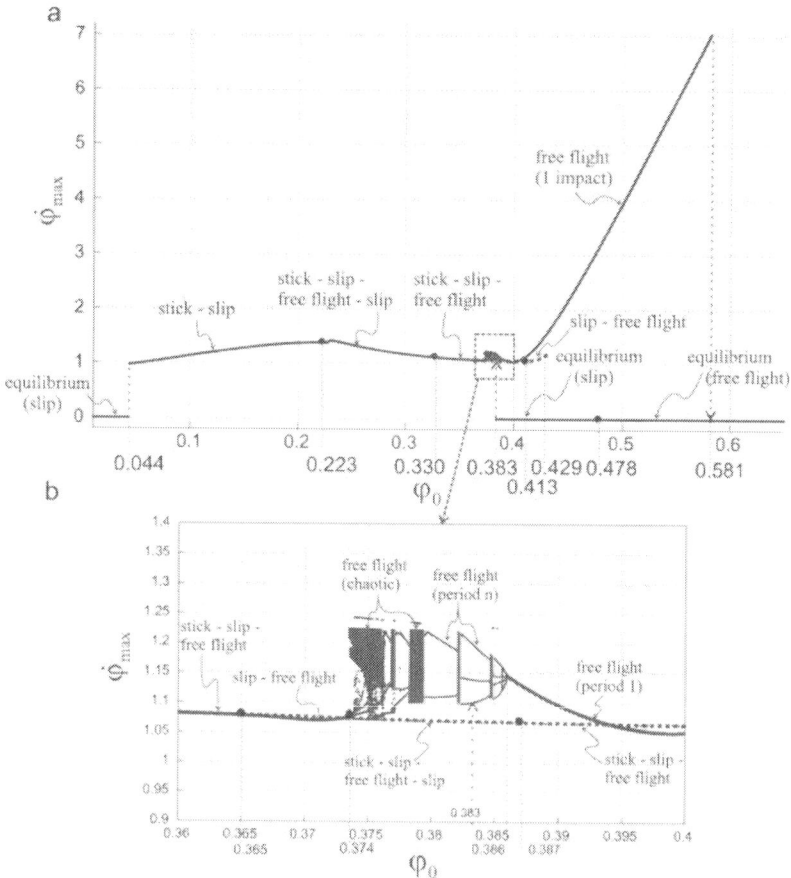

Figure 6. Simulation 1. (a) Bifurcation diagrams for varying ϕ_0. $\varepsilon N=0.2\equiv$ solid line; $\varepsilon N=0\equiv$ dashed line,(b) zoom of a portion of (a).

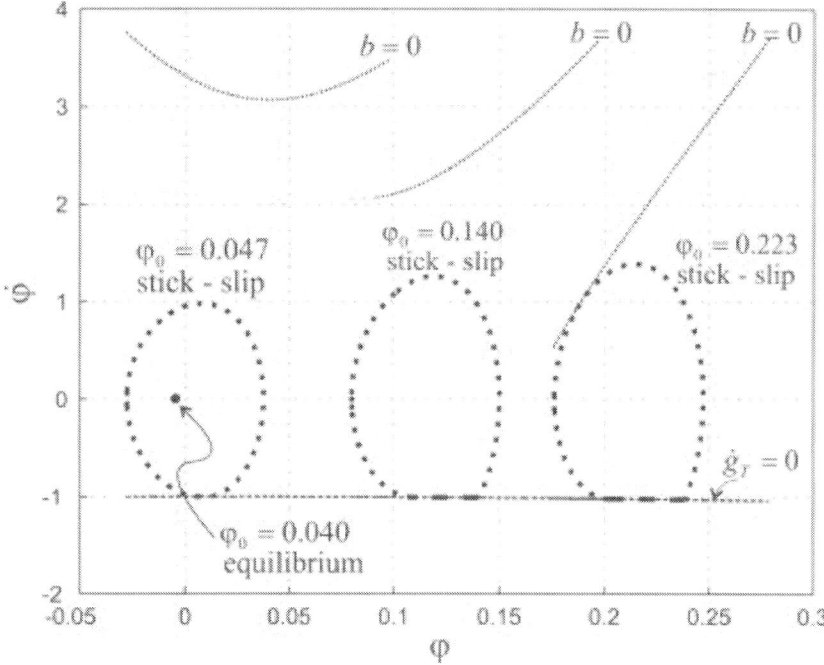

Figure 7. Simulation 1. Laid motions. Orbits in the $\varphi - \dot{\varphi}$ phase plane. Slip ≡ dotted line; stick ≡ dashedline.

When $\varepsilon_N = 0$ (dashed line in Figure. 6), the system exhibits a periodic motion of the stick–slip–free flight–slip type in the interval $0.223 < \phi_0 < 0.387$ (see the orbit for $\phi_0 = 0.350$ in Figure. 8(a)). In this range the post-impact slip phase reduces as ϕ_0 increases, and at $\phi_0 = 0.387$ the post-impact regime passes from slip to stick, as confirmed by the fact that the starting impact point $(\varphi_A, \dot{\varphi}_A, \dot{y}_A)$ satisfies Eq. (25)$H\dot{g}_{NA} = \dot{g}_{TA}$(this can be clearly seen in Figure. 8(b)).

In the interval $0.387 < \phi_0 < 0.413$, the periodic motion is stick–slip–free flight. The stick phase reduces as ϕ_0 increases and at $\phi_0 = 0.413$ it vanishes. The orbit at this transition point is represented in Figure. 8(c).

Looking to the enlargement of Figure. 8(c), we notice that the impact point $(\varphi_A, \dot{\varphi}_A)$ does not belong to the curve $H\dot{g}_{NA} = \dot{g}_{TA}$, even if the post-impact motion passes from stick to slip. The reason is that the impact problem is solved by a stick solution, i.e., $\dot{g}_{NE} = \dot{g}_{TE} = 0$ and the impulses Λ_{NE} and Λ_{TE}, evaluated by (17), satisfy the inequality

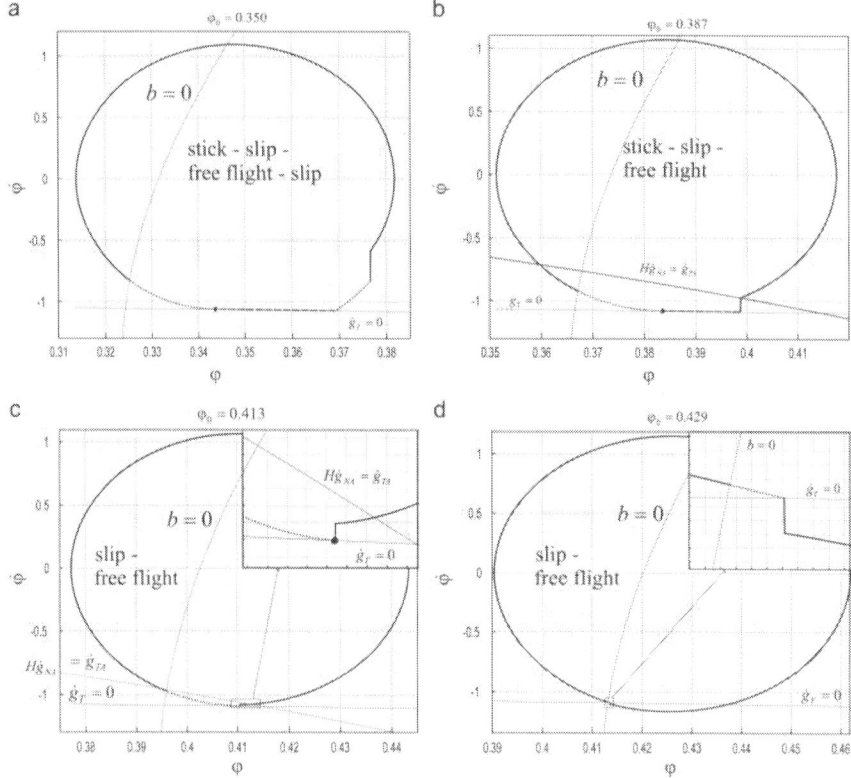

Figure 8. Simulation 1. Case of plastic impact ($\varepsilon_N=0$). Orbits in the $\varphi-\dot\varphi$ phase plane. Free-flight $\equiv$ solid line; slip $\equiv$ dotted line; stick $\equiv$ dashed line.

$|\lambda_{TE}|\leqslant\mu\lambda_{NE}$. But the tangential and normal forces at the instant $t=t_E$ (impact end) are such that $|\lambda_T(t_E)|>\mu\lambda_N(t_E)$. As a result, the stick mode cannot start, and only slip motion is allowed with the initial conditions $\phi=_{\phi A}$ and $\dot\varphi=\dot{s}/(l\cos\varphi_A)$. The slip mode starts with null relative tangential velocity. The absurd condition $|\lambda_T(t_E)|>\mu\lambda_N(t_E)$, which holds only at the initial instant of the slip post-impact phase, is due to the fact that, at impact,the Coulomb's law is extended to impulses and thus averaged forces are evaluated instead of punctual ones. This is a well-known feature of the considered model [21].

In the range $0.413<_{\phi 0}<0.429$, the motion has the slip and free-flight phases. Within a period, the duration of the slip phase reduces as $_{\phi 0}$ increases, and at $_{\phi 0}=0.429$ it vanishes (see Figure. 8(d)). At this point the orbit is all in free-flight regime, but an impact occurs. The solution of the impact law is a slip point, which belongs to the curve $b=0$. Thus, after the impact, the motion would continue in slip mode, however it cannot since the orbit enters the region $b>0$, where only the free-flight regime satisfies the complementary problem (37). As a result a post-impact motion does not exist and the bifurcation branch interrupts. This phenomenon is a consequence of the plastic impact assumption $\varepsilon_N=0$. In fact, it will no longer exist in the elasto-plastic case $\varepsilon_N=0.2$ analyzed below.

For $\phi_0>0.429$ only an equilibrium branch survives. The equilibrium is in slip mode for $0.429<\phi_0<0.478$ and in free-flight mode for $\phi_0>0.478$. The stable equilibrium branch starts at $\phi_0=0.383$. In the range $0.383<\phi_0<0.413$ the periodic and equilibrium branches coexist (hysteretic behavior).

Now we consider the case $\varepsilon_N=0.2$. Since for $\phi_0<0.223$ the mass m_1 never detaches from the belt, in this range the motion corresponds to that of the case $\varepsilon_N=0$. In the interval $0.223<\phi_0<0.330$ the bifurcation curves of the cases $\varepsilon_N=0.2$ and $\varepsilon_N=0$ practically coincides. In the case of elasto-plastic impact, the transition from free-flight to slip occurs through a 'chattering' phenomenon (see Section 3.2.1). Actually, only the first bounces can be seen in Figure. 9(a), due to the graphical resolution. The bifurcation diagram almost overlaps with that characterized by $\varepsilon_N=0$ since this phenomenon practically does not modify the values $\dot{\varphi}_{max}$ of the orbits.

At $\phi_0=0.330$, the post-impact slip phase vanishes (the corresponding orbit is represented in Figure. 9(b)). We notice that the same bifurcation point for the case $\varepsilon_N=0$ is $\phi_0=0.387$. In the range $0.330<\phi_0<0.365$ the periodic motion is of the type slip-free flight-stick. The stick phase reduces up to vanish at the point $\phi_0=0.365$ (see the orbit in Figure. 9(c)). For $\phi_0=0.365$ the 'chattering' finishes at the point of intersection between the friction cone and the curve $\dot{g}_T=0$. The corresponding bifurcation point for the case $\varepsilon_N=0$ is $\phi_0=0.413$. The slip phase vanishes at $\phi_0=0.374$ (see the phase portrait in Figure. 9(d)), and only the free-flight regime remains, with bounces between free-flight phases

We remind that in the case $\varepsilon_N=0$ orbits with only free-flight are impossible, and the bifurcation branch interrupts where they should appear (at $\phi_0=0.429$ in that case); here, on the contrary, we have shown that this kind of motion exists. More precisely, in the range $0.374<\phi_0<0.386$ the solution consists of chaotic bounces (see Figure. 9(e)) or periodic bounces of period $n>1$ (see Figure. 9(f) for a period 3 motion). The bouncing becomes of period 1 for $0.386<\phi_0<0.581$. The value of $\dot{\varphi}_{max}$ rapidly increases, as the bounces rapidly increase their amplitudes. The phase portrait for $\phi_0=0.5$ is represented in Figure. 9(g). The periodic branch ends at $\phi_0=0.581$ and, for $\phi_0>0.581$, only the equilibrium branch in free-flight mode survives.

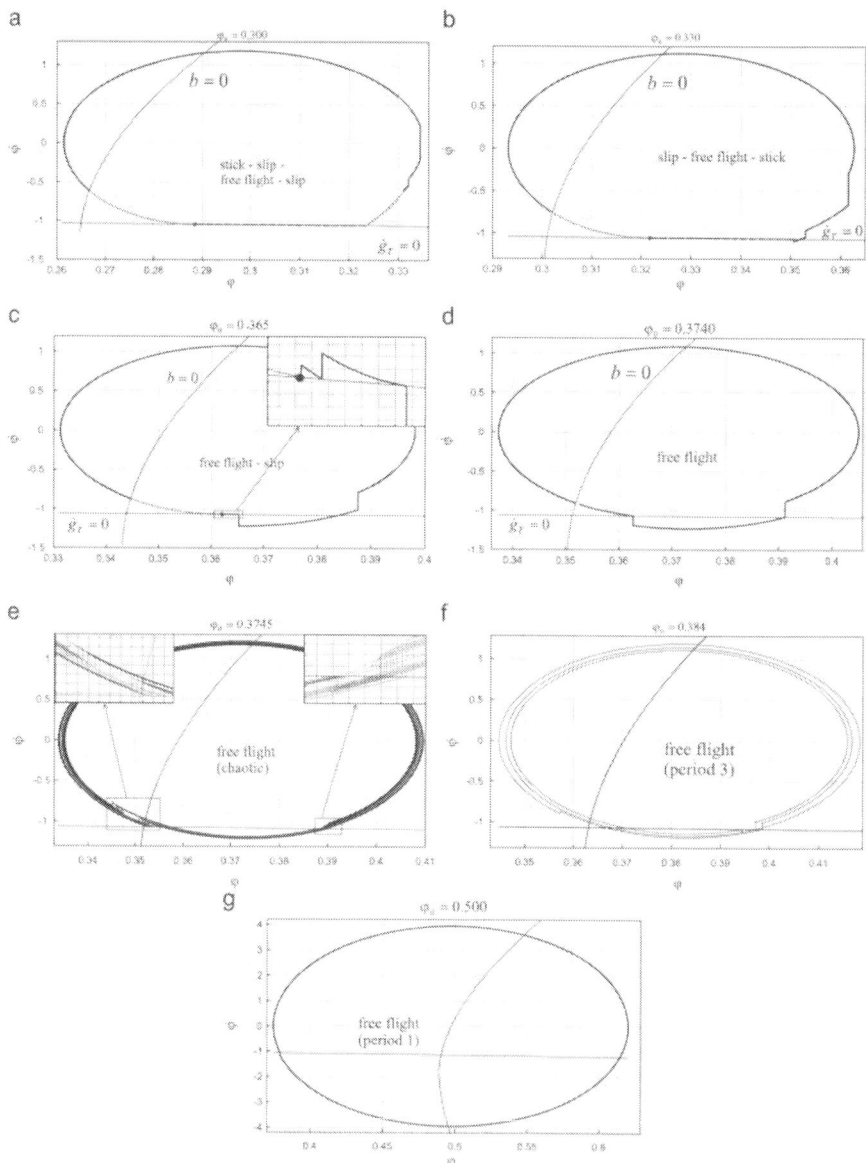

Figure 9. Simulation 1. Case of elasto-plastic impact ($\varepsilon_N = 0.2$). Orbits in the $\varphi - \dot{\varphi}$ phase plane. Free-flight ≡ solid line; slip ≡ dotted line; stick ≡ dashed line.

Simulation 2

We now assume $m_1 = 0.4\,\mathrm{kg}$ and $k_\varphi = 25\,\mathrm{Nm}$, a case in which the geometrical non-linearity plays a more important role on the dynamics.

Bifurcation Diagrams for Varying M

We fix $\phi_0 = \pi/8$. The bifurcation diagrams for varying μ are reported in Figure. 10. In this case, the diagrams obtained with $\varepsilon_N = 0.2$ and 0 *practically* coincide, also in the regions where there is free-flight. The bifurcation curve is constituted by four branches related to four different dynamical behaviors. The branch $\mu < 0.353$ is an equilibrium one. The next three branches correspond to periodic motions: for $0.353 < \mu < 0.360$ the motion is an oscillation in slip mode, for $0.360 < \mu < 0.453$ the periodic oscillation involves also the stick mode, and for $\mu > 0.453$ a more complex free flight–slip–stick–slip motion takes place. In Figure. 10(b) the bifurcation diagram is enlarged in correspondence of the slip oscillating branch. The transition from equilibrium to slip is a Hopf bifurcation occurring at $\mu = 0.353$, a value which coincides with that analytically evaluated in Section 4.1.

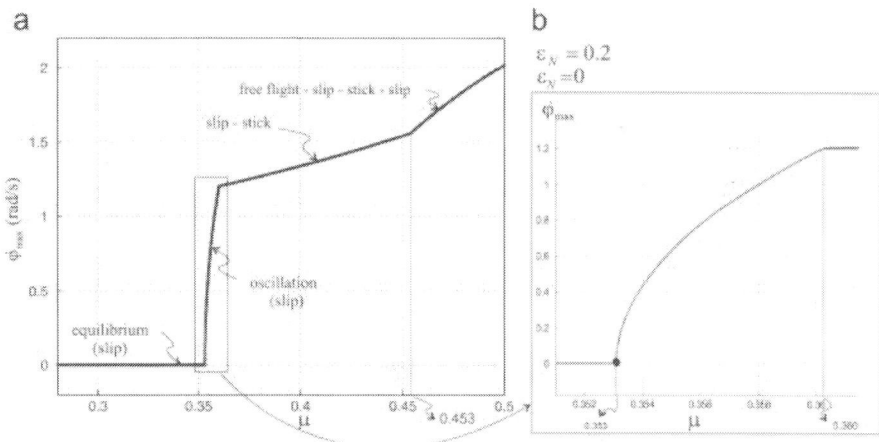

Figure 10. Simulation 2. Bifurcation diagrams for varying μ.

The orbits in the $(\phi, \dot{\phi})$ phase space are reported in Figure. 11 for different values of μ, where we observe that the larger is μ the larger is the diameter of the orbits. The bifurcation values $\mu = 0.360$ and 0.453 correspond to orbits which are tangent to the curves $\dot{\vartheta}_T = 0$ (appearance of a stick phase) and $b = 0$ (appearance of free-flight), respectively.

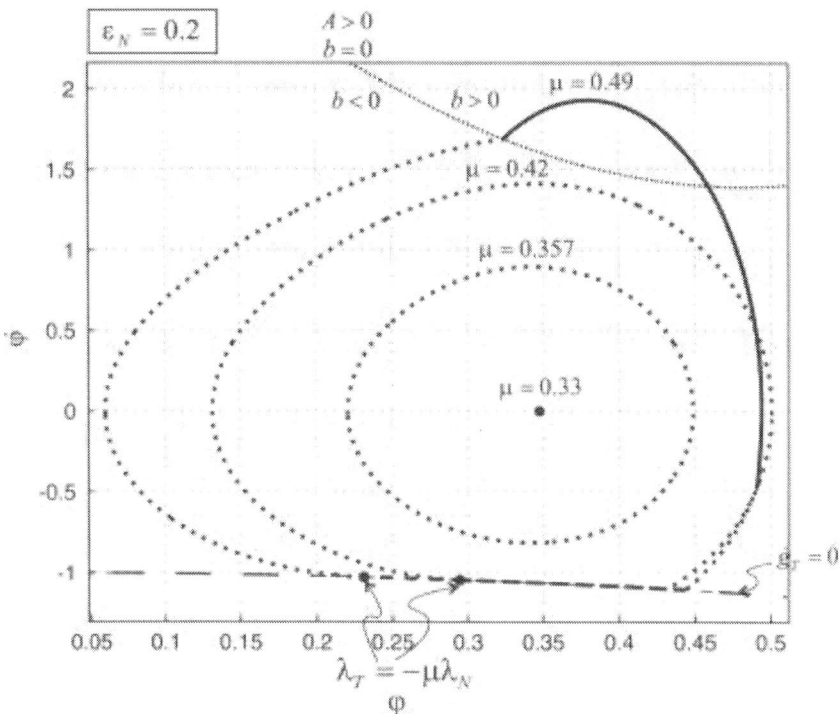

Figure 11. Simulation 2. Orbits in the $\varphi - \dot{\varphi}$ phase plane. Free-flight $\equiv$ solid line; slip $\equiv$ dottedline; stick $\equiv$ dashed line.

The motion experiences impacts only for $\mu>0.453$, and, as a result, only in this region the periodic motion depends on the impact law. Figure. 12 shows the orbits and the time histories of the m_1 vertical displacement for $\mu=0.49$, for both $\varepsilon N=0.2$ and 0. In the case $\varepsilon=0.2$, the solution goes from free-flight to slip through a cascade of accumulating impacts. On the contrary, in the case $\varepsilon N=0$, a single impact occurs between free-flight and slip. As shown in Figure. 12 the two orbits differ only in correspondence of the impacts and in the successive slip phase, and they practically coincide in the remaining part. This explains why the bifurcation diagrams for $\varepsilon N=0.2$ and 0 appear to coincide also when $\mu>0.453$.

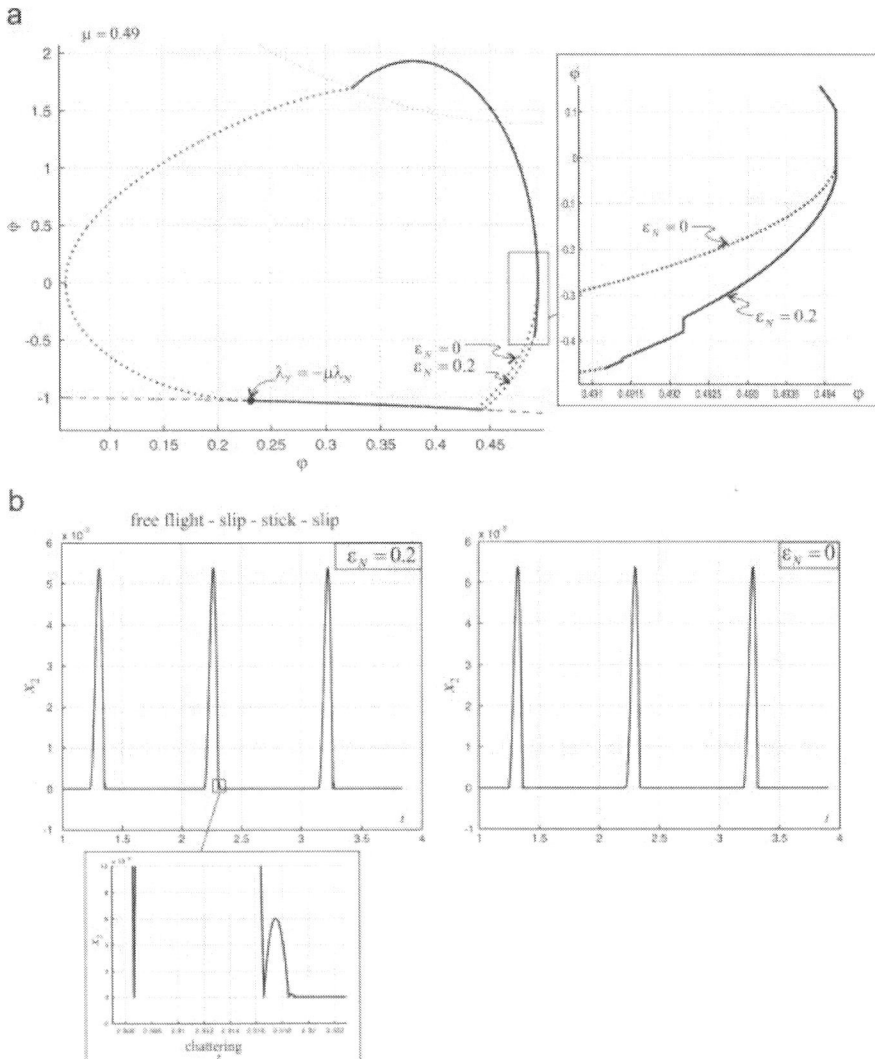

Figure 12. Simulation 2. Orbits in the $\varphi - \dot{\varphi}$ phase plane and time histories.

Bifurcation Diagrams for Varying ϕ_0

We now fix $\mu=0.4$ and vary ϕ_0. The bifurcation diagrams are represented in Figure. 13(a). Since the mass$_{m1}$ never detaches from the belt, the diagrams for $\varepsilon_N=0$ and $\varepsilon_N=0.2$ coincide. For $\phi_0<0.225$ and $\phi_0>0.442$ the system stays in equilibrium in slip mode. For $0.225<\phi_0<0.434$ the motion is periodic of the type stick–slip (see Figure. 13(b) for the orbit at $\phi_0=0.4$), and for $0.434<\phi_0<0.442$ the motion becomes periodic in slip mode. The orbit detaches from the curve $\dot{g}_T=0$, as shown in Figure. 13(c).

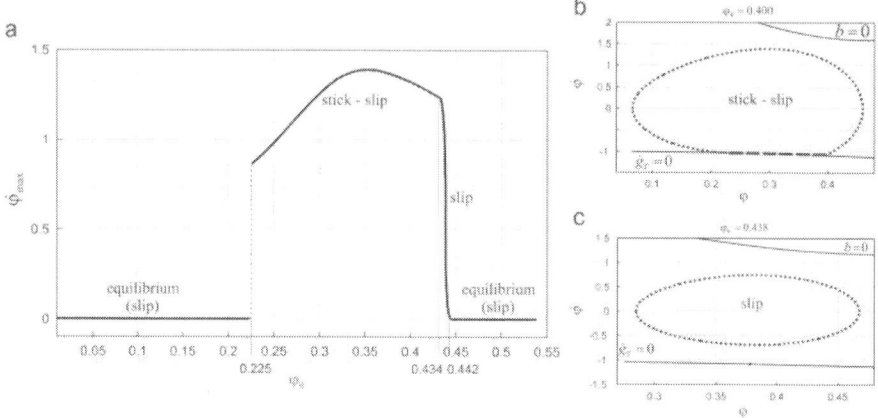

Figure 13. Simulation 2. (a) Bifurcation diagrams for varying ϕ_0; (b) and (c) orbits for fixed values of ϕ_0.

Surface Moving With Harmonic Velocity

In this subsection we assume a harmonic velocity for the moving surface:

$$\dot{s}(t) = A \sin\left(\frac{2\pi}{T}t\right)$$

43

And assign the values $A = 1\,\mathrm{m/s}$ and $T = 4\,\mathrm{s}$. All the other data are the same as in Simulation 1. In these preliminary tests we just vary the values of some parameters to see their effect on the dynamics. In particular, we analyze the influence of the parameters μ, ϕ_0 and y_0, focusing on the hopping phenomenon.

Influence of M

We fix $\phi_0 = \pi/8$ and $y_0 = 1\,\mathrm{m}$, and analyze the influence of the frictional coefficient μ. Varying the value of μ, it is found that mass m_1 never detaches from the belt if $\mu < 0.414$. The orbit in the $(\phi, \dot{\phi})$ space and the horizontal displacement of m_1 during a period T are plotted in Figure. 14(a) for $\mu = 0.4$. The mass m_1 mainly slips on the moving surface and sticks only when $\dot{s}$ changes sign. For $\dot{s} < 0$ the motion is an oscillation which asymptotically tends toward the point $\phi \approx 0.388$, while, for $\dot{s} > 0$, the oscillation quickly approaches the point $\phi \approx 0.409$. The asymmetry of the motion with respect to $\varphi = \dot{\varphi} = 0$ is due to the non-null unstressed conFigureuration of the rotational spring ($\phi_0 = \pi/8$).

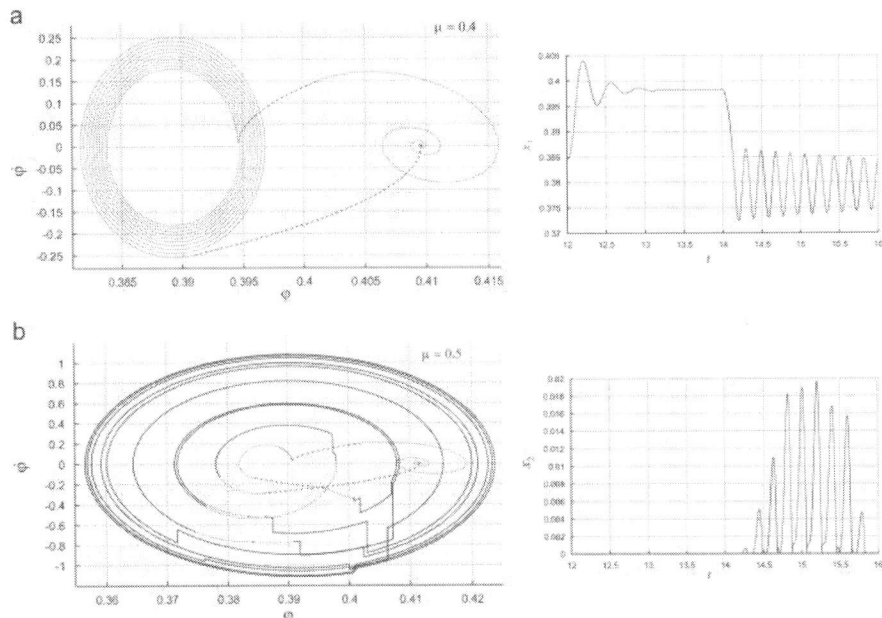

Figure 14. Sinusoidal surface velocity. Phase portraits and time histories for: (a) $\mu = 0.4$, (b) $\mu = 0.5$.

For $\mu > 0.414$, bounces take place in the dynamics. As shown in Figure. 14(b) for $\mu = 0.5$, the hopping occurs when $\dot{s} < 0$, and the motion is essentially in free-flight with short slip periods. When $\dot{s} > 0$, the motion is mainly an oscillation in slip mode. Stick mode happens in a time interval during which $\dot{s}$ changes sign.

The motions in the cases of elasto-plastic and perfectly plastic impact models are compared in Figure. 15 for $\mu = 1$. In both cases bounces occur in the half period $\dot{s} < 0$ and oscillations in slip mode take place when $\dot{s} > 0$. When $\varepsilon_N = 0$, stick phases are interposed within the free-flight phases. Looking at the time history of the vertical position x_2 of m_1 in the right part of Figure. 15, we notice that, in the case $\varepsilon_N = 0.2$, the first bounce in the second half period is the highest. Such a bounce is preceded by the stick phase which pumps potential energy in the system, later transformed into (released) kinematic energy during the subsequent free-flight phase. We notice that also the last bounce within the period, preceded by a stick phase, reaches a remarkable height. In the case $\varepsilon_N=0$, the bounces first increase and then decrease their height. The highest bounce happens when $\dot{s} \simeq -1$ (minimum value). In this case the amount of potential energy pumped in the system during the stick phases seems to be proportional to $|\dot{s}|$.

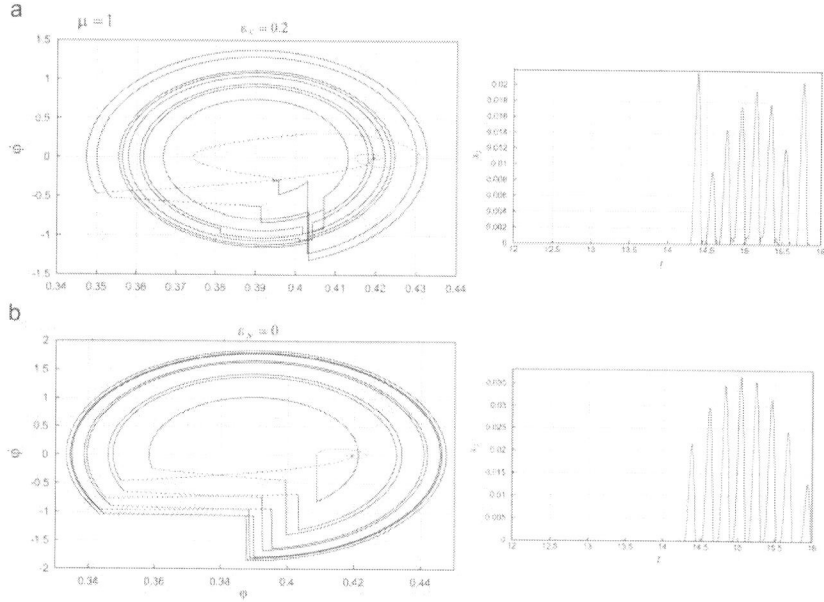

Figure 15. Sinusoidal surface velocity. Phase portraits and time histories for $\mu=1$: (a) $\varepsilon_N=0.2$, (b) $\varepsilon_N=0$.

These simulations reveal the well-known dependence of the hopping on the friction. A large value of the friction coefficient facilitates the hopping phenomenon. In fact, according to common sense, we know that, in the case of a windscreen wiper, the hopping occurs when the glass is dirty and dry, i.e., when μ has a large value.

Influence of ϕ_0
In the next simulations we test the dependence of the motion on the unstressed position of the rotational spring ϕ_0. We fix $\mu=0.4$ and $y_0 = 1\,\mathrm{m}$, and vary ϕ_0. Figure. 16 represents the motions for different values of ϕ_0 in the case $\varepsilon_N=0.2$.

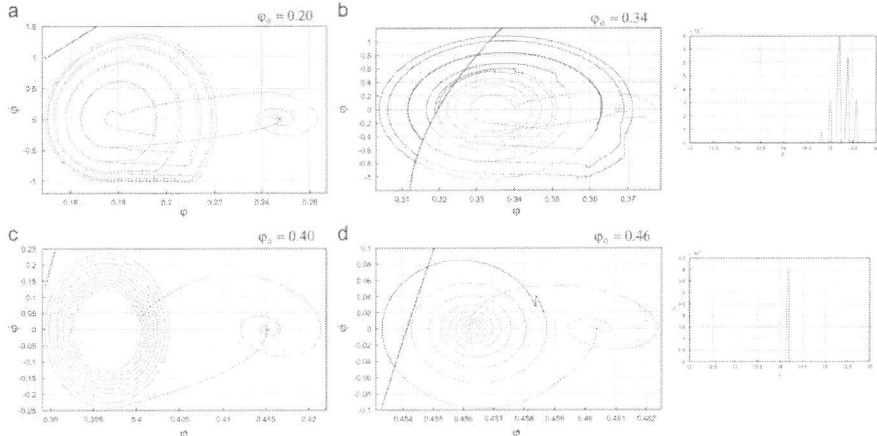

Figure 16. Sinusoidal surface velocity. Phase portraits and time histories for: (a) $\phi=0.20$, (b) $\phi=0.34$ (c) $\phi=0.40$, (d) $\phi=0.46$.$_{\varepsilon N}=0.2$.

The numerical tests reveal that for $_{\phi 0}<0.23$ the mass $_{m1}$ never detaches from the surface and the motion consists of oscillations which tend alternatively toward two accumulation points. The stick phase takes place when the velocity changes sign and also in the half period $\dot{s}<0$, alternating to the slip phase (see Figure. 16(a)). In the half period $\dot{s}>0$, the oscillation is in slip mode. For $_{\phi 0}>0.23$, the mass $_{m1}$ detaches from the moving surface. The motion exhibits free flight–slip–stick oscillations when $\dot{s}<0$ and slip oscillations when $\dot{s}>0$ (Figure. 16(b)). As $_{\phi 0}$ increases, the amplitudes of the oscillations reduce, and, consequently, the orbits occupy smaller and smaller regions in the $(\varphi,\ \dot{\varphi})$ plane (compare the length of the coordinate axes in the phase portraits of Figure. 16). The free-flight phase disappear when $0.375<_{\phi 0}<0.430$ (Figure. 16(c)). When $0.43<_{\phi 0}<0.48$ a small bounce per period takes place (Figure. 16(d)). For $_{\phi 0}>0.478$ (value previously found in the bifurcation diagrams of Figure. 6), the mass $_{m1}$ detaches from the surface and stays in equilibrium in free-flight.

Referring to the windscreen wiper application, these simulations suggest that the hopping phenomenon can depend on the permanent deformation of the wiper blade which affects the blade rubber when its elastic properties are compromised and irreversible plastic deformations occur. Therefore, the phenomenon should be more frequent with old wipers, affected by a larger 'permanent deformation'.

Influence of y_0

Finally, we change the value of the unstressed conFigureuration y_0 of the linear spring. We fix $\mu = 0.4$ and $\phi_0 = 0.34$. As shown in Figure. 17, if we reduces the value of y_0 to 0.924, the bounces reduce in number and amplitude. For $y_0 = 0.739$, the free-flight phases disappear. When $\dot{s} > 0$, slip oscillations quickly converge toward the point $\phi \approx 0.58$ and, when $\dot{s} < 0$, stick–slip oscillations take place.

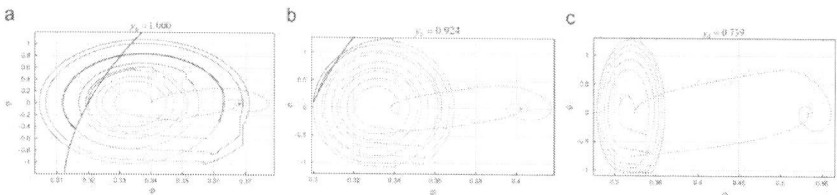

Figure 17. Sinusoidal surface velocity. Phase portraits for: (a) $y_0 = 1$, (b) $y_0 = 0.924$, (c) $y_0 = 0.739$.

These results show that, again in the interpretation of the model as windscreen wiper, the hopping depends on the wiper metal arm, and it increases when there is not enough (static) push on the rubber blade toward the glass.

CONCLUSIONS

A system with a varying number of dof affected by geometric non-linearity and non-smoothness due to friction and impacts is investigated. The paper highlights how an apparently simple mechanical model (i) requires a complex set of transition conditions to link phases of motion characterized by a varying number of governing equations and (ii) possesses, as a consequence, a very complex dynamics where many different dynamical phenomena have been observed and related to the properties of the system. Both the non-smooth and the varying-dimension characters have been shown to play a key role.

It is clarified that the adopted mathematical framework is not fully consistent—a price to be paid to its relative simplicity. In the numerical code we developed for the simulations we insert appropriate warnings when various inconsistencies (or paradoxes) occur. In the numerical simulations shown in the paper we encounter only one of them.

In the case of constant driving velocity of a moving surface, the system dynamics is investigated for varying values of both the friction coefficient (μ) and the unstressed position of the rotational spring ($_{\phi 0}$). The bifurcation diagrams generated with a restitution coefficient in the normal direction equal to zero depict a wide variety of motions, and some of them were not highlighted in the previous publications.

The introduction of a restitution coefficient in the normal direction different from zero allows a more realistic description of the system dynamics, accounting for bounces and a phenomenon in which the transition from free-flight to contact regime occurs through infinitely many impacts, named 'chattering'. Moreover, in the considered numerical simulations, in several situations this avoids the Painlevé paradox and the related jump of the reactive forces, and in some other cases it avoids the interruption of periodic bifurcation branches.

Finally, some simulations are presented for the case of sinusoidal velocity of the moving surface. These simple examples seem to justify the adoption of the model under investigation to study the hopping phenomenon in windscreen wipers, for which it was originally developed.

ACKNOWLEDGMENTS

This work has been developed in the framework of the international research project 'Dynamics and control of engineering systems affected by friction forces,' founded by the Royal Society of London. The authors thank Dr. Laura Marcheggiani for her help in the numerical simulations.

REFERENCES

1. M. di Bernardo, P. Kowalczyk, A. Nordmark, Bifurcations of dynamical systems with sliding: derivation of normal form mappings, Physica D 170 (2002) 175–205.
2. C. Budd, F. Dux, Chattering and related behaviour in impact oscillators, Philos. Trans. R. Soc. London 347 (1994) 365–389.
3. S.-C. Chang, H.-L. Pin, Chaos attitude motion and chaos control in an automotive wiper system, Int. J. Solids Struct. 41 (2004) 3491–3504.
4. H. Dankowicz, A. Nordmark, On the origin and bifurcations of stick–slip oscillations, Physica D 136 (2000) 280–302.

5. L. Demeio, S. Lenci, Asymptotic analysis of chattering oscillations for an impacting inverted pendulum, Q. J. Mech. Appl. Math. 59 (2006) 419–434.

6. S. Foale, S.R. Bishop, Dynamical complexities of forced impacting systems, Philos. Trans. R. Soc. London 338 (1992) 547–556.

7. U. Galvanetto, Some discontinuous bifurcations in a two-block stick–slip system, J. Sound Vibr. 248 (2001) 653–669.

8. S. Goto, H. Takahashi, T. Oya, Clarification of the mechanism of wiper blade rubber squeal noise generation, JSAE Rev. 22 (2001) 57–62.

9. N. Hinrichs, M. Oestreich, K. Popp, Dynamics of oscillators with impact andfriction, Chaos SolitonsFract. 8 (1997) 535–558.

10. J.B. Keller, Impact with friction, ASME J. Appl. Mech. 53 (1985) 1–4.

11. G. Lancioni, S. Lenci, U. Galvanetto, The dynamics of a two-DOF system constrained by a frictional mono-lateral surface, Dynamics and control of engineering systems affected by friction forces, Report of the Royal Society grant, 2007.

12. R.I. Leine, B. Brogliato, H. Nijmeijer, Periodic motion and bifurcations induced by the Painlevé paradox, Eur. J. Mech. A/Solids 21 (2002) 869–896.

13. R.I. Leine, D.H. van Campen, Bifurcation phenomena in non-smooth dynamical systems, Eur. J. Mech. A/Solids 25 (2006) 595–616.

14. R.I. Leine, H. Nijmeijer, Dynamics and bifurcations in non-smooth mechanical systems, Lecture Notes in Applied and Computational Mechanics, vol. 18, Springer, Berlin, Heidelberg, New York, 2004.

15. C. Liu, Z. Zhao, B. Chen, The bouncing motion appearing in a robotic system with unilateral constraint, Non-linear Dyn. 49 (2007) 217–232.

16. Nordmark, Non-periodic motion caused by grazing incidence in an impact oscillator, J. Sound Vibr. 145 (1991) 279–297.

17. Nordmark, H. Dankowicz, A. Champneys, Nonsmooth bifurcations in systems with impact and friction: I. Discontinuities in the impact law, 2008, preprint.

18. Nordmark, P. Piiroinen, Simulation and stability analysis of impacting systems with complete chattering, Non-linear Dyn. (2009), to appear.

19. S. Okura, Dynamic analysis of blade reversal behavior in a windshield wiper system, SAE Technical Paper Series 2001-01-0127, 2000.

20. E. Pavlovskaia, M. Wiercigroch, C. Grebogi, Modeling of an impact system with a drift, Phys. Rev. E 64 (2001).

21. F. Pfeiffer, C. Glocker, Multibody Dynamics with Unilateral Contacts, Wiley, New York, 1996.

22. F. Pfeiffer, M.O. Foerg, On the structure of multiple impact system, Non-linear Dyn. 42 (2005) 101–112.

23. R. Suzuki, K. Yasuda, Analysis of chatter vibration in an automotive wiper assembly, Int. J. Jpn. Soc. Mech. Eng. Ser. C 41 (1998) 616–620.

24. Z. Zhao, C. Liu, W. Ma, B. Chen, Experimental investigation of the Painlevéparadox in a robotic system, ASME J. Appl. Mech. 75 (2008) 041006.

CITATION

Giovanni Lancionia, , Stefano Lencia, UgoGalvanettob, Non-linear dynamics of a mechanical system with a frictional unilateral constraint, doi:10.1016/j. ijnonlinmec.2009.02.012

Index